RENEWABLE RESOURCES AND GLOBAL CHALLENGES

RENEWABLE RESOURCES AND GLOBAL CHALLENGES

Edited by:

Mohit Chhabra

Delve PUBLISHING

www.delvepublishing.com

Renewable Resources and Global Challenges

Mohit Chhabra

Delve Publishing
2010 Winston Park Drive,
2nd Floor
Oakville, ON L6H 5R7
Canada
www.delvepublishing.com
Tel: 001-289-291-7705
001-905-616-2116
Fax: 001-289-291-7601
Email: orders@arclereducation.com

ISBN: 978-1-77407-273-8 (Hardcover)

ABOUT THE EDITOR

Mohit Chhabra is an experienced professional in the power industry, having focused on power hardware in the loop and controller hardware in the loop testing, microgrid controller testing and validation, distributed energy resource integration challenges, and medium and high voltage short circuit testing. Mohit has Doctor of Philosophy, Master of Science, and Bachelor of Science degrees in Electrical Engineering.

TABLE OF CONTENTS

LIST OF FIGURES

LIST OF ABBREVIATIONS

CA	Copenhagen agreements
CFLs	compact fluorescent lamps
CO2	carbon dioxide
CO2E/kWh	carbon dioxide equivalent per kilowatt-hour
CSP	concentrated solar power
DER	distributed energy resource
DOE	department of energy
EIA	Energy Information Administration
EPA	Environmental Protection Agency
EROI	energy returns on investment
FFV	flexible fuel vehicles
FO	furnace oil
GEA	geothermal energy association
GHG	greenhouse gases
HDI	human development index
IEA	International Energy Agency
IED	intelligent electronic devices
IPPC	Intergovernmental Panel on Climate Change
IRENA	International Renewable Energy Agency
IUCN	International Union for the Conservation of Nature
LNG	liquid natural gas
MJ/kg	megajoules per kilogram
MWh	megawatt-hour
NREL	National Renewable Energy Laboratory
OLED	organic LED
PV	photovoltaic
SOS	security of supply
TW	terawatts

PREFACE

The resources that are generally used by the people for their basic needs and other tasks, may vary in a wide range. They have predominantly been non-renewable for quite a lot of time, which has resulted in a rapid depletion of such resources.

However, the people have been able to find better alternatives to such resources which can be found in renewable forms. These kinds of resources help the people to sustain their lives in a better and sustainable way. This book has tried to bring the focus of the readers to the various renewable forms of resources and the global challenges they can help in dealing with.

The book starts by discussing the biotic resources of energy. It explains the readers about the ways in which the living organisms contribute towards energy production in the environment and its usage. It takes the readers through the ways that humans have been using their resources.

Moving further, the book dwells on the issues related to the biotic resources. It emphasizes the need for distributing the resources equally and intelligently for their sustainable utilization. Then, it lays importance on the role of individuals in contributing to the conservation of nature. It discusses the future of natural resources in the energy sector.

The next chapters talk about the abiotic resources sun as the sun, water, wind, forests, geothermal energy, other such forms. They inform about the various ways in which these resources can be used to harness the energy and supply in the forms in which it can be utilized. The book also enlists various challenges that arise while using each of these resources of energy.

Further, the chapters go on to discuss various other renewable sources of energy. They give an insight into on the ways that these renewable resources can be used to generate electricity. The book details the procedures and equipment used to harness the energy and supply it in the form of electricity.

It goes on to discuss the alternative resources of renewable energy. It explains all the concepts and recent researches on the alternatives for renewable energy. It also enlists the differences between the alternative energy resources and renewable energy resources. It emphasizes on the need for such resources and the benefits that they provide in the long run.

Going further, the book talks about the non-renewable resources of energy. It informs the readers about the history of their usage. The readers are further informed about the various non-renewable forms from which energy can be extracted in various ways.

The book enlists several impacts of using non-renewable sources of energy. It includes social, environmental, and economic impacts. The book also dwells upon the future of non-renewable energy resources and the possibilities that exist in this field for them.

The book discusses about the technical advancements in the renewable energy sector. It talks about the prospects of various technologies and the possibilities that lie in the future in this industry. It emphasizes on the need to produce clean energy and the policies that are evolving keeping such an objective in mind.

1

Introduction to Renewable Resources

CONTENTS

Renewable sources of energy are eco-friendly sources of energy that are freely available and easily replenishable. These sources of energy are known to have a number of applications that are quite effective in energy conservation and achieving energy security in several nations. There is a need to understand the history of renewable sources of energy which has been explained in detail. The need for renewable energy and the different types of renewable sources of energy have been elucidated in detail. The different applications of renewable sources of energy and the future of the energy sector have been explained in detail.

1.1. INTRODUCTION

Renewable sources of energy refer to the sources of those energies that are naturally acquired from the atmosphere and that are utilized from the sources that are naturally replenished and available. There are different types of renewable energy sources such as solar energy, wind energy, geothermal energy, hydropower, and biomass.

Figure 1.1: Representation of renewable sources of energy.

Sources: https://cdn.pixabay.com/photo/2018/09/17/12/05/energy-3683673_960_720.jpg, https://cdn.pixabay.com/photo/2017/04/01/15/05/

big-waves-2193828_960_720.jpg, https://www.flickr.com/photos/ly-dur/4132279118, https://c.pxhere.com/photos/d0/41/wind_energy_renewable_enegy_windmill_wind_mill_energy_rotation_power-993970.jpg!d, https://cdn.pixabay.com/photo/2017/07/11/08/33/dam-2492809_960_720.jpg

Energy is something that is utilized everyday and is being used in day to day lives of people. The increase in science and technology and arrival of a number of novel inventions is leading to a number of gadgets and equipment that are put to use in everyday life. The increase in number of innovations that facilitate human beings lives everyday are known to demand and need enormous amounts of energy which is being used from fossil fuels.

However, the use of a number of fossil fuels such as propane or electricity for lighting, heating, and powering of our devices is known to have created a huge amount of need for a variety of fossil fuels in large quantities because these are depleting non-renewable sources of energy. Therefore, to meet the demands of the energy requirement people started becoming more and more dependent on the renewable sources of energy that are freely available.

There are a number of items such as computers, phone networks, servers, and security systems, shopping malls, parking lots, sports stadiums, cars, airplanes, and others which are increasing rapidly with the increase in technology. Majority of these aspects need power which is generated from fuel and this power or fuel is mostly generated either from renewable sources of energy or from fossil fuels, however, the utilization of renewable sources of energy is gradually increasing over time because of the various negative impacts of fossil fuels and the presence of limited resources of fuel.

People belonging to different countries and governments are working hard to reduce the carbon emissions across the world and improve the consequence that is rising because of the use of fossil fuels such as climate change and global warming.

There are different laws and conventions that are coming up for the promotion of the use of renewable sources of energy, and one such convention was the Paris Climate Summit which had a new agreement that was made in 2015 which is known to have made an agreement on the reduction of the global average temperature. Some of the important aspects of this convention that were made to increase the eco-friendly policies of the environment are as follows:

- To move forward, we also need to realize that there is only so much that can possibly be done in limiting GHG output as the human population only increases and puts more demands on our energy infrastructure.

- There is another aspect which is an important integral aspect of this convention which states that there is need to improve steps and work hard to improve and conserve the environment so that it becomes safe and secure for the future generations. Therefore, to make the planet safe and secure for the future generations, there is need to encourage the use of renewable sources of energy that are more eco-friendly without any negative impact on the environment.

1.2. A HISTORY OF RENEWABLE ENERGY

Coal deposits are known to have changed the face of technology and industries by becoming important sources of renewable energy; however, it may be extremely surprising to learn that during the period of industrial revolution before the discovery of coal deposits the major sources of energy for heating and lighting was mostly renewable sources. However, there were few exceptions with respect to the use of these renewable energy sources during that period.

However, the discovery of coal changed several aspects related to industrial revolution in the western world and later still learned to tap oil in greater quantities leading to an acceleration of technologies that would take us into the 20th century.

Plant material such as wood, grass, mosses, and others were burnt in the historic and prehistoric times to make food for several other activities. This process of burning such organic matter for energy sources is called biomass in the recent times. The process of burning this organic matter became the central aspect and major source of energy in recent time because of which hearth and fireplace because this organic matter is known to be the major source of energy.

There is a perception of several people, which suggests that the discovery and utilization of fire in the history of civilization marks the beginning of the use of renewable energy. Human civilization and development of humanity continued evolving in a similar way from several thousands of years, which began even before the discovery of oils. The initial discovery of oil for using as sources of energy initially happened several centuries ago, and very small traces of oil were discovered and used. The use of technology in mass drilling of oil and exploration of several sources of oil has started in the industrial age. The other uses of renewable energies with respect to antiquity refer to the animal power, which indicated the use of cattle to drive plows or turn millstones and the use of wind for sailing purposes. These were practices that were implemented in trade nearly 8000 years back even before the beginning of the human history. The use of water sources, such as creating dams to harness the power of the fluid motion of water, is not a new idea either.

These ancient technologies were carefully researched, and several novel technologies were generated to supply power from renewable sources in the future. The concepts of peak oil and peak coal were made into theories in the historic times that were made in 1870s. There were concepts that developed and research which was made on several renewable sources of energy such as solar energy even during the period of industrial revolution.

These thoughts processes evolved over a period of time and started developing technologies to equip future generations to a post-coal world. Though the reasons to look for renewable sources and thoughts regarding these concepts related to renewable sources of energy might have changed over a period of time, but the need remains the same. There is a need to look for renewable sources of energy to prepare the future generations towards a post-oil world.

There is a clear-cut indication from the long time where there may be a possibility for the coal reserves and oil reserves to run out and create a shortage of energy supply for the growing population. There were a number of theories that originated during the World War I, and there is a huge investment in solar technology that was made during the same period. There was a research paper that was published in the year 1912 in Scientific American which is known to have made a hypothesis regarding the possible run out of fossil fuels which would lead a number of people with solar energy as the only option.

The arrival of the concept of the peak oil in the 1950s led to the development of a novel search for the renewable sources of energy. The different naturally available renewable sources of energy are solar energy, hydroelectricity, tidal power, and others which were explored by several environmentalists and industrialists. Though there was no shortage and reduction in the supply of oil or fossil fuels, they were often regarded as finite sources that may become extinct at any point irrespective of the quantity and frequency of availability. There was an increase in the awareness regarding several environments and there were measures that were taken to reduce the pollution (such as the Clean Air Act in the US and equivalents in other countries most of which passed in the 1960s–1970s) that is being created because of the fossil fuels as a result of which the awareness and use of renewable sources of energy were not just innovations of science, but they were the necessity for giving a safe and healthy planet for future generations.

There are several arguments and discussions regarding the reach of peak oil and peak coal; however, there were many people who believed that peak oil phase occurred in 2008. There was an interesting finding around the same period that the availability of novel pockets of oil was becoming scarce, and there were limited opportunities to find new sources of oil.

There was a steep and sharp increase in the demand for oil ever since 1986 which made several scholars and environmentalists preach about the possible hazards of fossil fuels and look for stable sources of energy such as renewable sources of energy to meet energy demands of the growing population. There has been a steady instability in the production of oil in several non-oil producing countries which led to the arrival of a novel concept called energy security.

In addition to several factors, the major issue related to energy security was an important issue of concern for many leaders and scholars across the world by the end of the 20th century and the beginning of the 21st century. Energy security refers to the interrelationship between every country's national security and the presence of several resources in that country with respect to energy production and consumption.

The presence of limited resources or restricted access to the resources of oil and energy in any particular nation is known to lead to a huge amount of instability which may cause political disturbance and armed conflict among nations. Energy security and lack of sufficient resources for energy could be a major war causing factor for the future generations as oil is a major commodity. Therefore, to ensure energy security across several nations, it is necessary to explore several alternate sources of energy such as renewable sources of energy which are freely available and can be easily replenished.

1.3. NEED FOR RENEWABLE ENERGY

1.3.1. Fossil Fuels Are Limited

One of the important reasons for the utilization of renewable sources of energy is that fossil fuels are very less in terms of availability. Fossil fuels are popularly called as finite sources of energy as they are present in limited quantities at places that are extremely scarce in terms of availability. These fossil fuels are also called as non-renewable sources of energy because they cannot be naturally replenished.

There is a huge scarcity of fossil fuels that is happening, and day by day, there is depletion in the availability and utilization of fossil fuels. However, it must be understood that though the availability of fossil fuels has not reached a peak oil which is a stage where the demand outstrips supply, the current situation indicated that in the future there is a possibility to reach this situation very soon, because of which there is need for the conservation of resources and explore for new sources of energy to meet the growing demand of the people.

In addition to the existing scenario there is a situation that is created by several external factors that play an important role in creation of a huge demand for these natural resources which may lead the way for an acute shortage in the future as a result of there is an active research for more and more ways for exploration and utilization of several renewable sources of energy that is being created.

Because the fossil fuels are depleting and the demand for energy across the globe is increasing there is an active exploration that is started by governments and businesses to look for a greater number of options and sources for exploitation and utilization of renewable sources of energy. These factors and existing situation indicate and represents a scenario where fossil fuels will eventually fade over time, and there would be the arrival of several sources and uses for the renewable sources of energy that is known to change the phase of the entire market in the upcoming years.

Another important aspect of the fossil fuels is that it takes nearly 10,000,000 years for the production of fossil fuels that have been used up in 150 years as a result of which it becomes extremely difficult to meet the increasing energy demands of the ever-increasing population across the globe.

The human population is dynamic and is increasing constantly over a period of time, and with the increase in population, there is an increase in the demand for the energy which can be achieved by tapping the existing natural sources that are freely available, easy to use and can be easily replenished.

The major constraint with respect to the fossil fuels is that there are a number of challenges that are faced by several geologists over a period of time to explore the sources of crude oil. Geologists state that there is a steady increase in the difficulty with respect to the exploration and location of sources for crude oil.

The main area if attention is not to concentrate on the availability of the fossil fuels or the quantity of the fossil fuels that are left, but the major area of focus is to look for ways to exploit and utilize the available natural sources of renewable energy that are both eco-friendly and sustainable.

1.3.2. Carbon Emissions and Climate Change

The major problem and immediate consequence of the use of fossil fuels are the carbon emissions that are released by fossil fuels that led to climate change. There was a COP21 agreement that was made in 2016 in which there was stress on the climate change and the carbon emissions that were released from fossil fuels that had extremely hazardous impact on the climate over the past 150 years.

The release of carbon emissions from the fossil fuels was known to have had a huge impact on the climate and environment as a result of which several environmentalists have campaigned and tried to bring awareness about the use of fossil fuels and carbon emissions.

There were drastic weather changes, global warming, and climate change that have had an impact on the entire world with changing the climate. There were different continents and several countries that have recorded extremely high temperatures in summer and record low temperatures in the winters in the last few years. In addition to these drastic changes in climate, there were other environmental changes such as typhoons and hurricanes, dry spells, droughts, and flooding.

These extreme fluctuations in weather were recorded in all the countries, and they have affected every country. These changes in weather were largely attributed to the increasing levels of pollution, which is because of the increase in carbon emission. Fossil fuels were known to be the major contributors to the carbon emissions leading to further increase in pollution.

Therefore, renewable sources of energy were identified as eco-friendly and low carbon emission sources of energy which can become potential sources of energy for the future. These renewable sources of energy are known to have extremely little or absolutely no carbon output, and they are capable of catering several energy needs of individuals. However, the possibility to become completely carbon neutral is difficult as there would be carbon output when making a solar panel or building a dam and similar other sources.

Therefore, there is need to take stringent steps that must be implemented and followed to reduce the carbon footprint for several international regulations to help those people in the developing world to protect themselves from the crazy climate. There is another important aspect related to climate change such as melting of ice caps and the rise of sea levels, which may create food shortage and result in national instability. Therefore, renewable energy sources are being exploited and utilized because they are cheap, easily available, and naturally replenished sources of energy.

1.3.3. Energy Security

Energy security is known to be a major need for exploration of different renewable sources of energy and using these renewable sources of energy. There has been a lot of instability in many areas in the Middle East with respect to the various sources of energy. The major reason for a lot of economic and political instability in the Middle East is because of energy.

Middle Eastern countries are known to be reservoirs of oil, and they are the largest suppliers of the oil to various parts of the world. There are different resource of energy in different parts of the country such as South America is known to produce oil, North America and South America are known to famous for the supply of coal and several other European countries, United Kingdom and Russia are known to house for mining gas. New tension between Russia and the west, firstly over Ukraine and secondly over Syria, has led to increased distrust between world powers.

There are some countries which are huge suppliers of fossil fuels, and there are some countries which do not have any supply of any energy source. This creates huge interdependency among nations for energy, which also becomes an important source for the creation of relationships or leads to sour relationships among nations promoting them to have armed conflict. This interdependency for energy may lead to a lot of conflicts among nations because energy is one factor that would have a huge impact on the economy of a nation resulting in destabilization of the economy of a nation. Therefore, to avoid this situation, there is a need to equip a nation with various sources of renewable energy so that the nation can work towards energy security and independence of energy sources to avoid any conflicts.

Energy security is an important aspect of any nation as it must be understood that fossil fuels are finite sources of energy and there is an increasing demand in energy with the increase in the population which further strengthens the need for energy security. If a nation is not self-sufficient with respect to the sources of energy, then there is a possibility that the nation initially becomes economically unstable, which further leads to political instability. It is expected that increased tension over acquisition and protection of resources could lead to global conflict.

There are some experts who state that the major crisis in Syria is mainly because of the energy security issues and in addition, there are some more people who perceive that the possibility of next world war could be large because of energy dependency. However, an important way to overcome these barriers is trying to explore for alternate sources of renewable energy such as solar energy, wind energy, geothermal energy, hydroelectricity, and several others which can meet the demands of increasing population and result in ensuring energy security across the world.

There is a huge difference and fluctuation of oil in the past 10–15 years, which ranged from an extremely high price in 2012 to 2013 to an extremely low price record in 2015–2016. Oil and energy are the major drivers or barriers for any economy and environment as a result of which such huge fluctuation in the price of oil led to a number of protests and strikes because of these highs and lows in prices.

There is a need to recognize that oil is an important commodity with a huge possibility for depletion, and when these prices are erratic and changing there is a need for exploration of different alternative sources of energy.

1.3.4. Economic Stability

The major possible achievement of the renewable sources such as hydroelectricity, tidal power, solar panels, and others is the constant availability of these resources, and they are economically viable and sustainable. Renewable sources of energy are known to create economic sustainability as there would be no price fluctuations in maintenance and use of these resources. The energy that is derived from renewable sources is known to be extremely cheap when compared to non-renewable sources or fossil fuels. This is the major reason for the economic sustainability that is achieved from the utilization of these resources.

A number of countries and places use a major chunk of energy from several renewable sources of energy such as geothermal sources of energy. There are a number of places such as Texas where the energy produced from renewable energy sources such as wind energy is made available to citizens at an extremely cheap rate. Therefore, these renewable sources of energy are known to create a lot of economic stability in several places and countries.

1.3.5. Environmental Damage

Fossil fuels are extremely difficult to procure as there are limited resources for fossil fuels and because there are limited resources for fossil fuels, the prospectors start searching new areas and pockets for oil and other fossil fuels. The need to look for new resources for fossil fuels, which will make these prospectors look for several areas and continuously drill at different locations because of which there would be enormous amounts of environmental pollution. In addition, because of the presence of limited resources of oil and other fossil fuels, there would be drilling that would happen for longer depths to get the oil.

This drilling practice for prospecting oil and fossil fuels would lead to a lot of disturbance in the environment and ecosystem and create a lot of problems to the flora and fauna that are living around these areas. The pollution caused because of this is the main reason for the debate regarding the several issues, and there is a continuous conflict between several governments, industries, and environmental groups regarding the impact of continuous drilling and prospecting for oil on the environment. There are a number of forests, aquatic flora and fauna, and other wildlife that is being affected because of the environment.

A number of developed countries and the people living in these countries are known to be affected the most because of the hazards caused by environmental damage and pollution. There are a number of protests and strikes that are happening in several countries against the drilling and fossil fuel prospecting problems that are arising.

Therefore, because the fossil fuels are leading a huge amount of environmental ruckus, there is a need for an alternative source of energy because of which people are showing interest and exploring the possibility of utilization of freely available natural sources of energy to meet their needs for energy.

Renewable sources of energy are known to be sustainable green sources of energy that have very little negative impact on the environment and they are known to be extremely important eco-friendly sources of energy because of which there are different governments and people who are trying to increase the use of renewable energies in daily life and reduce the utilization of fossil fuels that are harmful to the environment.

1.3.6. Public Health

The non-renewable sources of energy such as oil, gas, coal drilling and mining are known to create enormous quantities of pollution because of which there are a number of side effects that have come up in the local environments and the areas surrounding them.

There are a number of strikes and protests that are being carried out by several people with respect to the presence of fossil fuels and their negative impact on the health of the people because of which there are many people who are stressing on the importance of use of alternative sources of energy instead of the fossil fuels which or non-renewable sources of energy that have a lot of negative impact on the health of people and environment. The carbon emission that is released by fossil fuels and the environmental hazards that are created by the use of fossil fuels have a lot of negative impact on the life of people.

The renewable sources of energy are environmentally friendly and mostly emission free because of which they do not create any impact on the life of the individuals. The most important aspect is that the majority of renewable sources of energy are known to be environmentally friendly and do not create any impact on the health because of which they are used and favored by several governments and people.

1.4. TYPES OF RENEWABLE ENERGY

There are majorly five types of renewable energy sources that are obtained from the environment and are widely used for different sustainable forms of energy. The different types of renewable energy are as follows:

- Solar energy;
- Wind energy;
- Geothermal;

- Hydroelectricity;
- Tidal power;
- Alternative sources of energy, such as bioenergy and hydrogen.

1.4.1. Solar Energy

Solar energy is the most tapped and utilized sources of renewable energy that is used every day. Solar energy refers to the energy that is acquired from the sun. There are several technologies that are being developed based on the energy that is acquired from the sun, and it is later converted into heat. Solar energy is being used for illumination, hot water, electricity, and several other cooling systems in several industries and businesses (Figure 1.2).

Figure 1.2: Representation of solar energy.

Source:https://www.maxpixel.net/static/photo/1x/Light-Sun-Person-Solar-Energy-Pear-Light-Bulb-3065254.jpg

There are some novel technologies such as photovoltaic (PV) systems that are being developed to convert the sunlight into electricity. There are a number of huge buildings that use solar water heaters that circulate water through flat-plate solar collectors. There are certain mirrored dishes that concentrate on boiling water in a traditional steam generator that is capable of concentrating sun heat and generate electricity. There are a number of shopping complexes, commercial buildings, and other industries that are utilizing solar energy as an alternative source of energy for different needs, such as ventilation, heating, and cooling. In addition to all these, there are several architectural designs that are using solar energy for different forms of heating and cooling.

Solar energy is exploited by several homeowners, business, and government entities for applications in different areas of life. The different walks of life in which solar energy is being used in different areas of life are as follows:

- Installation of a home solar system and commercial solar panels for residential purposes.
- In the construction of a building that would need solar panels which would be able to utilize solar energy for different hot water, cooling, or ventilation systems.
- There are several people who are capable of designing structures that are capable of utilizing the different energy sources that are important for heating and cooling systems.

1.4.2. Wind Energy

Wind energy is the energy that is generally created from winds, and the winds are created generally because of the presence of uneven heating and cooling of the atmosphere, which is known to create winds. An additional factor that is known to contribute to the creation of winds is the rotation of the earth and several other topological factors that lead to the formation of wind. Wind energy is sometimes considered as solar energy because of the uneven heating and cooling of the atmosphere that is known to result in the formation of winds.

The flow the winds are generally captured by the wind turbines and these wind turbines are known to convert the entire energy into electricity. However, on smaller scales, windmills are used to pump water on several small-scale farms (Figure 1.3).

Wind turbines are available in both small scale and large scale. The large-scale wind turbines are known to be of commercial grade wind turbines. These commercial grade wind turbines are known to be efficient and professional made with the incorporation of several wind-powered engineering technologies which are capable of becoming efficient sources of renewable energy for several commercial organizations.

Figure 1.3: Representation of wind energy.

Source: https://www.maxpixel.net/static/photo/1x/Wind-Turbine-Power-Wind-Turbine-Energy-Electricity-954509.jpg

The wind turbines are categorized into two broad categories, and they are as follows:

- Single-wind turbines; and
- Utility-scale wind farms.

Single-wind turbines are those turbines that are known to produce electricity and assist the power supply to an existing system of electricity. When the wind blows, the power generated by the system goes to offset the need for utility-supplied electricity. Utility-scale wind farms are those that are made to produce electricity that can be bought and utilized by a wholesale power market. This electricity from the utility-scale wind farms can either be bough contractually or through bidding amidst a regular competition.

1.4.3. Geothermal: Power from the Earth

Geothermal energy is known to be a form of energy that is obtained from the heat that is present on the surface of the earth. The source for the geothermal source of energy would generally be the heat that is present close to the surface or from the heated rock and reservoirs of the hot water miles that might be present deep below the surface of the earth.

The primary purpose of these geothermal power plants is that these power plants are generally made to produce electricity from this heat. On a much smaller scale, a geothermal heat pump system can leverage the constant temperature of the ground found just 10 feet under the surface to help supply heat to a nearby building in the winter or to help cool it in the summer (Figure 1.4).

Figure 1.4: Representation of geothermal power.

Source: https://upload.wikimedia.org/wikipedia/commons/thumb/3/3b/Krafla_geothermal_power_station_wiki.jpg/800px-Krafla_geothermal_power_station_wiki.jpg

Geothermal energy is a form of energy which is generally an integral part of a commercial utility energy solution on a bigger magnitude, or it could be a smaller part of the sustainable practice of energy at a local level. There are a number of areas in which geothermal energy is used, and there are several applications of geothermal energy among which the direct applications of geothermal energy are as follows:

- The important use of geothermal energy is to heat office buildings and manufacturing plants.
- Geothermal energy is used to grow greenhouse plants and is used in heating water at several fish farms.
- Geothermal energy is used in several industries for a number of industrial processes such as pasteurization of milk.

Therefore, it is understood that geothermal energy is an important type of geothermal energy which has a wide number of applications and is known to play an important role in meeting the energy demands of the people and ensuring energy security to the growing population.

1.4.4. Hydroelectricity

Hydroelectricity is not a novel invention; hydropower is known to be a renewable source of energy which was used from historic times. There were several ancient technologies such as waterwheels that were used in the operation of gristmills and sawmills that used the power of water. These technologies are present in several museums, and the use of hydroelectricity from ancient times is indicated in several historic sites and museums.

Kinetic energy refers to the energy that is gained by a moving body, and the water that is flowing is known to gain kinetic energy. This kinetic energy that is acquired by several flowing rivers is known to be used in several ways and is converted to electricity. An important way by which the hydroelectric power is utilized is produced by a system where there are several dams that were constructed to store water in a reservoir and utilize turbines to generate electricity from this water.

There are majorly two methods by which water is generated, and they are as follows:

- Pumped storage hydropower; and
- Run-of-river hydropower.

1.4.4.1. Pumped Storage Hydropower

This is a type of reservoir where water is cycled between two reservoirs called the upper reservoir and the lower reservoir. This cycling of water between upper and lower reservoirs is known to control the production of electricity during the times of peak and low demands.

1.4.4.2. Run-of-River Hydropower

The other method of generation of electricity is the "Run-of-River Hydropower" which is known to create a funnel portion in the flow of a river through a channel. This method does not need any dam.

There are different types of hydropower plants that are used in the generation of electricity, and there are big massive projects related to the hydroelectricity such as Hoover Dam and several micro-hydroelectric power systems that are used in the generation of electricity.

There are chances for direct utilization of hydroelectric power, and they are generally dependent on the geographic location. There are several ways that are being implemented in the construction and utilization of electricity in small farms and other setups for several micro-hydroelectric plants, and they are known to easily accessible and available. The small waterways that are present in several small towns are used by these small towns in the building of some medium-sized hydroelectric power systems.

1.4.5. Power from the Ocean

The ocean is a dynamic source of water and a huge powerhouse of energy. The ocean is used in multiple ways to produce energy, and there are several types of energy that are produced from the ocean, and they are broadly categorized into two types, and they are as following:

- The thermal energy that is produced from the heat of the sun; and
- Mechanical energy that is provided from the movement of tides and waves, which is called tidal energy.

1.4.5.1. Thermal Energy from Ocean

The heat energy from the ocean is used to convert energy from the surface water to electricity with the help of certain sophisticated technologies.

1.4.5.2. Ocean Mechanical Energy

The energy that is created from the rise and fall of tides is utilized in this form of energy. The tides in the ocean are largely created because of the rotation of the earth and the gravitational influence of the moon. The other source which plays an important role in the creation of tides in the ocean is known to be the wind energy that generates tides. These energy sources play an important role in the reduction of electricity costs by several magnitudes (Figure 1.5).

Figure 1.5: Power from ocean.

Source: https://images.pexels.com/photos/164234/pexels-photo-164234.jpeg?cs=srgb&dl=beach-ocean-sea-164234.jpg&fm=jpg

Generally sophisticated technologies are required for the exploitation and utilization of the energy from the oceans in any form; however, there are certain lesser-developed technologies that can utilize the tides, ocean currents and winds in the ocean in the generation of power.

The ocean water that is present at great depths is known to be extremely cold, and this cold water is known to be used in cooling the different building. The residential colonies and housing communities that are present along the side of the ocean are known to utilize the energy from the ocean in several municipal power and energy needs.

Ocean energy is an evolving source of alternative energy production, and with more than 70% of the surface of our planet covered by ocean, its future looks promising, depending on geographies and regulatory guidelines.

1.5. OTHER ALTERNATIVE SOURCES OF ENERGY

There are broadly two types of renewable energy which is being produced by the use of several mechanical means instead of the use of any natural process directly. There are two alternative sources of energy, and they are as follows:

- Bioenergy; and
- Hydrogen energy.

1.5.1. Bioenergy

Bioenergy is known to be a renewable source of energy that is obtained from the biomass. Biomass which is obtained from the organic matter, is utilized to create heat and electricity or to manufacture liquid fuels such as ethanol and biodiesel is known to be an important form of renewable energy.

Biomass is an organic matter that is obtained from several living plants and animals. The bioenergy that is produced from biomass is known to generate some carbon emissions. The carbon dioxide that is produced by the utilization of bioenergy is known to be similar in quantity to the amount of carbon dioxide that is generated from fossil fuels. However, the advantage with the bioenergy is that the plants that are grown for production of biomass are known to eliminate the same amount of carbon dioxide from the atmosphere that is being produced from the burning of these fuels. There are a variety of systems used to generate this type of electricity, ranging from directly burning biomass to capturing and using methane gas produced by the natural decomposition of organic material.

1.5.1.1. Uses of Bioenergy

Bioenergy is utilized in several businesses and organizations that are involved in the transportation of people or goods. These organizations use biofuels such as ethanol or biodiesel in their vehicles.

- There are several manufacturing facilities that can burn the biomass and generate the steam that is captured by turbines to generate electricity.
- Small towns can utilize the methane gas that is created by the anaerobic digestion of organic waste in landfills and utilize them as a fuel to generate electricity.
- Small scale farms can produce electricity from the wastes

produced by the livestock into electricity by the utilization of small modular systems that can convert the waste into electricity.

1.5.2. Hydrogen: High Energy/Low Pollution

Hydrogen is known to be the simple element that is available abundantly in the universe. Though, Hydrogen is present abundantly in the universe, it is not present as natural gas on the earth. Hydrogen is not present as a gas on the earth, but it is present in several organic compounds such as gasoline, methanol, propane, and natural gas. The other form in which Hydrogen is abundantly present on the planet earth is in the form of water. Hydrogen gas is produced by several algae and bacteria with the utilization of the light from the sun.

The importance of hydrogen is that hydrogen is known to be extremely high in energy and do not produce any pollution when it is burned. This is an alternative renewable source of energy which was initially utilized in launching space shuttles and other rockets into the orbit during the 1950s.

The fuel cells that are made out of hydrogen are known to convert the potential chemical energy of hydrogen into electricity. This process leads to water and fuel as by-products. These hydrogen fuel cells cannot be used as a source of energy on a commercial scale because of the exorbitant costs that are involved in the utilization of these cells as a source of energy.

There are some countries that use the Hydrogen, and they use hydrogen in refining petroleum, produce fertilizer, and treat metals and process foods. The hydrogen fuel cells are used in limited numbers as sources of energy, and they are used as sources of energy where hydrogen and oxygen atoms are combined to generate electricity. There are certain hydrogen-powered vehicles that are present in a few countries such as the United States; however, these can be utilized only with the reduction of costs that are involved in the production of these cells.

There is a number of other applications for the utilization of hydrogen as a renewable source of energy, and they are as follows:

- Large fuel cells that are responsible for providing emergency electricity for several buildings and remote locations.
- Electric motor vehicles powered by hydrogen fuel cells and marine vessels powered by hydrogen fuel cells.

1.6. APPLICATIONS OF RENEWABLE ENERGY

There are a number of applications of renewable energy, and these applications of renewable energy can be broadly classified into two types, and they are as follows:

- On-grid; and
- Off-grid.

A grid is basically an integration of generation, transmission, and distribution system which supplies energy to several consumers. These terminologies on-grid and off-grid are commonly used to describe the channel in which electricity is delivered.

- **On-Grid:** These systems generally deal directly with the power stations that are directly connected to grids such as wind farms and solar panels.
- **Off-Grid:** These applications are those that are generally connected to a small home or a village house which is generally single load.

There are several different applications of off-grid systems as they are present in several forms such as PV modules which are connected to a small home in a village to centralized windmills to power a village water pump or a commercial battery charging facility. The off-grid applications are generally present in a rural village setup, and the on-grid applications are generally useful in the generation of electricity in huge quantities.

There are several applications of the renewable sources of energy, and they are as follows:

- Wind energy is known to be present in huge quantities, and the wind turbines are known to convert huge winds into mechanical power, which is, in turn, converted into electricity. This wind energy is utilized to charge batteries and to pump water from borewells. There are several winds powered vehicles that are being developed which can utilize the energy of wind.
- Solar energy is another abundantly present easy form of renewable energy which is utilized in several applications such as to power PV panels that are an excellent source of generating electricity for small households. These are extremely reliable as they have low maintenance, and these are great sources of electricity in rural areas. Solar energy is widely used in the solar water heater,

calculators, and solar lights and in certain offices where glass PV modules are used for a steady supply of electricity. The solar energy panels work by storing energy from the sun during day and utilization of this energy at the night.

- Geothermal energy is that energy that is derived from the earth, and it is mostly utilized by farmers. They use this energy to heat their greenhouses, which enable them to grow various fruits and vegetables all around the year.
- There are some countries that utilize the geothermal energy to heat the pedestrian walkways and bicycle lanes that are used to prevent these pathways from freezing in extreme winters.

1.7. THE FUTURE OF ENERGY

The important aspect of ensuring energy security is the proper utilization of energy, which plays an important role in meeting the demand for energy by several people. There are several experts and scholars across the world who states that there is a need for exploitation and utilization of renewable sources of energy to ensure eco-friendly and safe sources of energy for future generations.

There is a need to increase the utilization of renewable sources of energy for the generation of power to meet the increasing demand of people for energy. The need to utilize fossil fuels to meet the demands of the growing population does not exist in the current situation and there no more fluctuation in the prices of fuel that might lead to economic and political instability across the country.

The current situation focuses on the utilization of safe and sustainable renewable sources of energy which are naturally available free of cost, and they are easily replenishable. Therefore, these natural renewable sources of energy can meet the demands of the growing population and would ensure energy security in the future.

The different types of renewable sources of energy that are freely available are hydropower, energy from oceans, solar energy, wind energy, geothermal energy and other sources of energy that can ensure sustainability and can create energy security across different countries in the world. The renewable sources of energy would provide scope for people to install solar panels on their rooftops and utilize these sources of energy for their homes.

There are nearly 18 different sources of renewable energy that are known to be eco-friendly, and they do not create any pollution. Therefore, these renewable sources of energy are known to be the future sources of energy across the planet. These renewable sources of energy are known to be much more economical than several fossil fuels, and they are responsible for not creation of any adverse effects on the environment.

Fossil fuels are known to damage the environment by several carbon emissions that create pollution whereas renewable sources of energy are known to be much more environmentally friendly because there are neither mining and drilling practices nor carbon emissions that can cause pollution in these renewable sources of energy. In addition to the several advantages of the renewable sources of energy, it must be understood that renewable energy sources are a lot more economically viable than fossil fuels. The future of energy in any country across the globe is regarding the conservation of energy, and successful exploitation of several renewable sources of energy is the future.

There are several vehicles that use gasoline as a source of gasoline and gasoline is known to be a fossil fuel. Gasoline is a fossil fuel which has finite resources, and there are chances that gasoline might deplete any day. Therefore, there is the need for vehicles and automobile companies to generate hybrid systems which can utilize both fossil fuels and renewable sources of energy as sources of power.

There are a number of ways to conserve power, and they are as follows:

- An important way to conserve power is by switching off lights in that room where there is no need for light. There is a need to minimize the utilization of lights when the sun is there, and it is an absolute necessity to use sunlight instead of burn electricity. This small practice of switching off lights when not present in that room will save a lot of energy and electricity bill.
- Incandescent lamps can be used instead of Compact Fluorescent Lamps (CFLs) as the incandescent lamps are known to consume less power when compared to the CFLs. The incandescent lamps are made in such a way that they consume limited power and give bright light. This would not just be economical for the household or industry, but it can lead to a lot of conservation of energy from others.
- The major contributor to the electricity bill in these days is the air-conditioning and heating systems that lead to enormous amounts

of consumption of electricity and might lead to wastage of energy. There is a need to adjust the thermostat of the air conditioner so that there is optimum use of electricity and there is an effective conservation of energy. For instance, if the heater is typically set to 71-degree Fahrenheit in the peak winter, set it to, for instance, 69 and if the air conditioner is at 73 in the peak summer, set it to, for instance, 75.

These small steps may lead to a huge difference in the environment and can lead to a lot of energy conservation. Conservation of energy, optimum utilization of electricity, and utilization of renewable sources of energy is the future of the energy sector.

It is a matter of fact to state that there are a number of people who are becoming extremely conscious of the importance of the utilization of renewable sources of energy, however, it must be understood that there is need to do a lot of work in this area and there is need for development of several technologies that can lead to the further utilization of different renewable sources of energy. There are several awareness programs and campaigns that are being done to increase awareness among people regarding several renewable sources of energy. There are several engineers and scientists who are working vigorously to develop effective technologies in the sector of renewable energies to effectively utilize the freely available renewable sources of energy.

There are a number of awareness campaigns regarding the hazards related to the utilization of fossil fuels and a negative impact that is created on the environment because of several fossil fuels. There is a need to make the courses related to renewable energy compulsory in schools and colleges so that children learn the importance of several renewable energy sources and the importance of conservation of energy.

There is a need for governments to revise their policies related to energy and make new policies that enable the complete utilization of renewable sources of energy. There is a need for novel technologies and innovations related to the energy sector that can solve the problem of an energy crisis and ensure energy security. There is a need for several cross-country technology exchange programs that must be initiated so that the developed countries can help several developing countries to establish, build, and reinforce the growth of the renewable energy sector.

1.8. CONCLUSION

Fossil fuels are known to be finite sources of energy that would deplete over a period of time, and they are known to have a negative impact on the environment by creating pollution and release of carbon emissions. There is a steady increase in the population and along with the increase in population; there is an increase in the need for energy.

Energy is the major requirement for several people because of several technologies and vehicles that are increasing every day. Energy could be the major reason for economic and political instability as a majority of countries are energy dependent. Therefore, to avoid any conflict among nations, there is a need to ensure energy independence in nations which are mostly achieved by tapping several renewable sources of energy.

Renewable sources of energy such as solar, wind, hydroelectricity, ocean, and others are known to be widely used renewable sources. Renewable sources of energy are eco-friendly, do not lead to any carbon emissions, cost-effective, and easily replenished as a result of which governments and public are exploring and utilizing these natural resources of energy for their benefit. There is a need to bring about awareness among people by schools and social media campaigns by these valuable sources of energy and gradually equip towards port oil and post-coal-free world.

REFERENCES

1. Davison, A., (2019). *Renewable Energy, forms and Types of Renewable Energy*. [online] Altenergy.org. Available at: http://www.altenergy.org/renewables/renewables.html (Accessed on 21 May 2019).
2. Science, E., (2019). *Renewable Energy: All You Need to Know*. [online] Environmentalscience.org. Available at: https://www.environmentalscience.org/renewable-energy (Accessed on 21 May 2019).
3. Shahzad, U., (2012). *The Need for Renewable Energy Sources*. [online] Iteejournal.org. Available at: http://www.iteejournal.org/Download_August15_pdf_4.pdf (Accessed on 21 May 2019).
4. SunPower (2019). *Learn About Seven Types of Renewable Energy | Sun Power Solar Blog*. [online] SunPower – United States. Available at: https://us.sunpower.com/blog/2018/02/23/learn-about-seven-types-renewable-energy (Accessed on 21 May 2019).

2

Biotic Resources and Its Global Challenges

CONTENTS

Biotic resources are defined as those resources that are obtained from living things. This mainly comprises coal, gas, petroleum, etc. which are generally obtained from plants. It eventually includes plants and animals. A great number of issues or challenges are also linked with the utilization of biotic resources that have been discussed in the chapter. The chapter initially talks about the biotic resources and the history of the consumption of these resources. Further, the associated issues are described along with the global perspective of these resources. Global challenges and the role of humans in the conservation of biotic resources are briefly described. At the end of the chapter, the future scenario of the biotic resources is briefly described.

2.1. INTRODUCTION TO BIOTIC RESOURCES

Biotic resources are defined as those resources, which are mainly obtained from the biosphere. It consists of various kinds of raw materials upon which the human life, as well as economic production, depend, the capacity of absorption that keeps the population from being suffocated in their own waste and the ecological services that further develops a habitat, which is capable of providing support to human life (Figure 2.1).

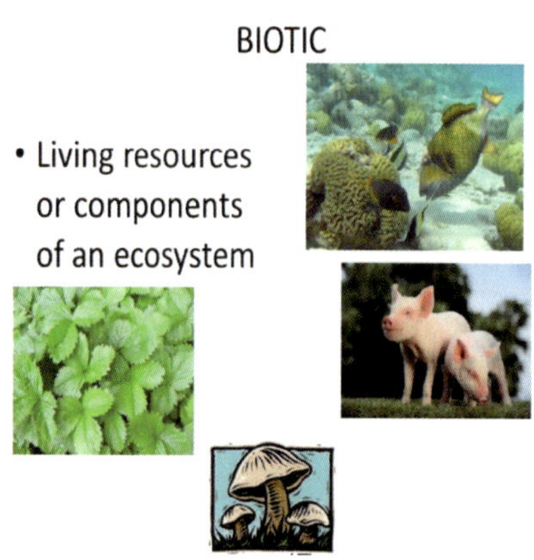

Figure 2.1: Biotic resources.

Source: https://slideplayer.com/slide/8564998/

As the non-renewable resources are being exhausted, human society is coming to depend more upon the self-renewing ability of the biotic resources.

Therefore, it is crucially important that people understand the nature of such kind of resources.

As the focus is turned to the biotic resources from abiotic ones, people must address a significant increase in the complexity, certainly accompanied by a critical increase in the uncertainty as well as ignorance.

One specific level of complexity ascends from the intrinsic value, which is provided, to the living systems. Abiotic resources are totally considered means to distinct ends, where one of the primary ends is the life sustenance, the maintenance of biotic resources.

Biotic resources enhance or improve the well-being of humans directly as well as they are considered an end in their own right, specifically so far responsive creatures. The biotic resources are physically complicated in two major ways.

- First is that the processes that are responsible for the persistent reproduction of individuals, populations, or species are very complex as well as poorly understood; and
- Second is that the individuals, populations, and species interact with other individuals, populations, and species, along with the abiotic resources to create or develop an ecosystem.

Ecosystems are extremely dynamic and complex, changing over time in essentially impulsive ways. Biotic resources are also regarded as living resources. They are defined as a renewable resource and comprise plants and animals.

Various products such as food, medicines, etc. are obtained from biotic resources, i.e., plants. The timber is obtained from the tree, which is a biotic resource and is utilized in the process of construction, making of ships, furniture, etc.

The wood pulp obtained from trees is used to make different kinds of paper. The wool, gums, alcohol, acetic acid, dry fruits, etc. are also obtained from the trees. They provide shelter to various other organisms. In addition, they provide the most basic element of oxygen to living beings. They severely affect the weather of a specific area.

The plants that are present on the land are generally classified as forests, shrubs, croplands, and grasslands. The vegetation greatly affects as it prevents soil erosion. The hummus is made by the decayed leaves, which further provide sufficient nutrition to the soil. In addition, the plants check the dessert spread. In a given year, the yearly production of the biomass mainly produced by marine plants is around 325 million tonnes. The overall production of biomass is around 380 million tonnes.

For survival, humans mainly depend on land vegetation. The earth has 4 million species of flora, out of which few are edible. Therefore, the conservation of the plants is a necessity. Deforestation, pollution in the environment, the formation of cropland and overgrazing causes a hostile effect on the growth and development of plants.

Few of the plant species are endangered. Thus, there must be a kind of check on the harmful activities of human beings. The marine plants can be utilized for numerous purposes. Humans should always explore those plants that are known as the good source of food.

In addition to that, the role of animals is very significant in the biotic environment. They are regarded as food for humans. The food is provided by the fish, prawns, sheep, goat, honey-bees, etc. some of them are also used as a mode of transport.

They consist of horses and mules, camel, asses, etc. also; some of them are utilized as a mode of agricultural processes. It includes camel and cattle.

Some of them are also used as the scavenger, which includes pigs and jackals mainly. Whereas, some of them are used as pearls which comprise the mollusks. The animals are known for maintaining the ecosystem. The fecal matter obtained from a few animals is utilized as manure.

The elephant and beer provide the ivory and musk. The Lac insect and honeybees respectively provide the Lac and wax. The process of pollination, as well as the dispersal of fruits and seeds, is specifically done by the insects. Few of these animals are also used in several laboratories as a research model (rats, rabbits, guinea pigs, etc.).

The secretion of hormones of few mammals treats the diseases of human beings, and hence the conservation of animals is needed. Poaching and killing of the animals, pollution, and deforestation cause a harmful effect on the growth of various animals.

Some of the animal's species are endangered. Therefore, there must be a check on human activities. In the 25 years from now, almost one-fourth of the animal's species are going to extinct. The people should have clear knowledge about the wild-life of animals.

Microorganisms, as the name itself, recommends consists of very small organisms that cannot be seen with naked eyes. They are present everywhere on the surface of the earth. The first person who discovered them was Anton van Leeuwenhoek.

It mainly consists of bacteria, fungi, virus, and protozoa. The fungi comprise the yeast and molds. Bacteria are present everywhere in the space, and their size is non-static. The size usually ranges from 0.1 micrometers to 750 micrometers.

The bacterial cells greatly differ in the shape and do not have any definite or precise structure. They can be bacilli, coccus, and spirillum stalked in shape. The fungi are those kinds of organisms that do not have chlorophyll. They are heterotrophic in nature. They are known as eukaryotic and a vascular organism (Figure 2.2).

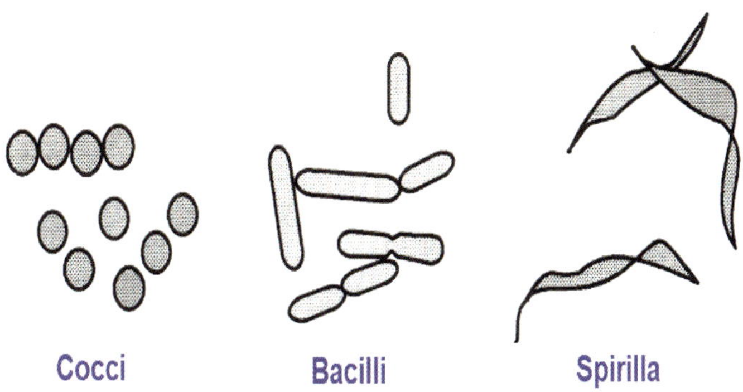

Figure 2.2: Different shapes of bacteria.

Source: https://commons.wikimedia.org/wiki/File:Bacteria_shapes_01.png

The cell wall of fungi is made up of chitin. Usually, they store their food in the glycogen form. A medicine, which is known as penicillin, is made from the fungi. This medicine is mainly used for treating the infections caused by bacteria. Fungi act as a saprophyte also.

They can create numerous problems too. The protozoa can behave as a beneficial as well as harmful organism. They sometimes act to decrease the fertility of soil pollute water to cause diseases. In addition, they act as a good source of food for aquatic animals.

The oil deposits are mainly found in the link with the skeletal deposits of radiolarian and other organisms. They aid to locate the fuel. More or less, all of the viruses are harmful or detrimental.

2.2. THE HISTORY OF HUMAN RESOURCE USE

Biotic resources have always been the basic material for the ecosystem. In human history, however, the per capita level of consumption of resources dramatically changed. At present, the inhabitant of the industrialized nations uses around four to eight times more resources as compared to those people who are living in the agricultural societies, and fifteen to thirty times more resources as compared to those who are living in hunter-gatherer societies.

Attaining a viable level of resource use universally does not mean that people should go back to the age of stone. Although, new or modern models are required to be found out that assure a good quality of life for every single person existing on the planet.

In human history, societies were always hooked on the utilization of biotic resources. Different kinds of societies dramatically consumed different level or amount of biotic resources. From the period of Stone Age until today, the per capita consumption of the biotic resources has literally grown by a factor of 15 to 30.

The cultural development of the humankind is the history of an ever-increased misuse of biotic resources. Early social systems like those of the hunter-gatherers and early agrarian societies were largely dependent upon the utilization of the renewable natural resources such as the sun and wood.

A hunter-gatherer society had a per capita consumption of natural resources of around one metric ton per person per year, which is equal to around 3 kg per day.

These resources were mainly used for basic food, housing, transportation, and weapons for bringing down the prey. In the case of agrarian societies, consumption increased to around four metric tonnes per person per year, which is equal to around 11 kg per day.

A huge percentage of this increase was because of the feed required for the animals, which were kept mainly for milk and meat production. At the same time, bigger buildings were established, and more metallic objects, such as weapons, plows, and cooking pots, were produced. These societies depended largely on wood as the key source of energy.

As one hectare of the forest could produce a limited amount of wood every year, the available energy to these societies was very limited. The economy, as well as the population, depend on the energy a lot and hence faced a limit to growth and development.

The industrial revolution during the 18th century proposed the most significant change in the use of biotic resources to date. With the utilization of fossil fuels such as coal, oil, and gas, more amount of energy was available to the humankind suddenly. The utilization of these fossil fuels, which were produced millions of years ago, developed a deceptive energy surplus for human societies.

This surplus was the major requirement for the economic growth that has sustained until today. It was the energy availability in a cheaper as well as more concentrated form that further allowed a sharp rise in the production of different kinds of goods and services.

Since the industrial revolution, the population has steadily grown, mainly due to the utilization of more machines and even more amount of the fertilizer aided a constant growth in the harvest yields per hectare of the cultivated land.

Hence, one hectare of the arable land at present feed a huge number of people as compared to the case in agrarian societies, which were restricted to the use of animal and human labor, and had no chemical fertilizers.

However, the progress or growth has had an immense environmental price, as resource consumption has dramatically increased. Today, a resident of an industrialized nation uses almost 15 to 30 tonnes of the raw materials and annually produces a multi-fold rise as compared to the agrarian societies.

At present, hunter-gatherer, as well as agrarian societies, still exists on the planet. Hunter-gatherer societies are usually found in the rainforests of Amazonia and Papua New Guinea.

However, these kinds of societies have been eliminated. The agricultural societies generally make up huge parts of the global South, in Asia, Africa, and Latin America.

However, a big share of the global population has moved into industrial societies as well as urban lifestyles. With a rapidly increasing global population, economy, and wealth, the consumption of nature also grows.

It would definitely help the population tremendously if global ecosystems would accordingly grow, but the size of the planet cannot be changed. How people can become more susceptible without going back to the period of Stone Age and without renouncing the ease of modern life?

In a world, which is regularly facing the physical limits, alternative ways, are required to be found out for the wellbeing and development of humans. Discovering new models of resource utilization is one of the major cornerstones of the modern approach of development. The major challenge is to assure a high quality of life for the current population of the world, which is around 7 billion, without surpassing the environmental capacities of the planet.

2.3. ECOSYSTEM STRUCTURE AND FUNCTION

An ecosystem is defined as the biological community of the interacting organisms along with their physical environment. It is a complex and interconnected system having a great community of living organisms as well as their non-living components. Nutrient cycles and energy flows are usually used to link the biotic and abiotic components of the ecosystem.

Various ecologists consider ecosystems in the context of structure and function, equivalent to the two kinds of physical complexities defined in previous section. This difference is very much related to economic analysis.

The economics of conventional natural resource is mainly the economics of the ecosystem structure. The environmental economics majorly focuses on assured functions of the ecosystem. As a matter of fact, structure, and function are interdependent mutually, and economics is required that efficiently integrates both.

Certainly, people must understand the interactions and differences between these two if these have to be incorporated into the economic analysis.

The structure of an ecosystem usually refers to the individuals as well as communities of animals and plants of which an ecosystem is made of, their age and spatial distribution and mainly the abiotic resources. Various ecosystems have a huge number of structural elements, each showing a variable degree of complexity.

Scientists have learned that when a sufficient number of separate elements are together thrown into a complex system, a kind of impulsive order results. One main property of such kind of system is their affinity to generate or produce emergent phenomenon, which can be defined as the properties of the whole that could not be assumed from a considerate of the individual parts.

The complex systems are usually categorized by the great nonlinear behavior, which means that a person cannot presume the results of huge interventions based on just an understanding of the smaller ones. For instance, removing a species stock of 40% from an ecosystem may have a distinct effect as compared to removing just 20%.

The structural elements function together in an ecosystem so as to create a whole, which is much greater than the sum of the parts (Figure 2.3).

Figure 2.3: A view of the ecosystem.

Source: https://www.nps.gov/hale/learn/nature/naturalfeaturesandecosystems.htm

In the ecosystem, these emergent phenomena are referred to as ecosystem functions, and they comprise various things such as nutrient cycling, energy transfer, climate regulation, water recycle, and the gas regulation.

These are typical of the evolving properties; ecosystem functions cannot be freely elucidated by even the widest knowledge of system components. Ignorance, uncertainty, and variability play a very significant role in the ecosystem structure analysis and a major role in the ecosystem function analysis.

People have a very limited and rare understanding of how the functions of ecosystem evolve from the complex connections of the ecosystem structure and hence, a tough time forecasting and sustaining the effects of human activities on such functions.

Thus, a major deal of uncertainty joins the process of decision-making consisting of the ecosystem functions. How people opt to treat uncertainty in the economic analysis is eventually an ethical or normative decision, yet an additional source of complexity.

One of the significant issues or problems related to the analysis of biotic resources is the amount of uncertainty involved. Solid instances always aid in clarifying any kind of concepts.

To demonstrate the connections between the structure and function along with the inferences of complexity, people must concentrate on a wet tropical forest, the terrestrial ecosystem that displays the extreme biodiversity of any yet studied.

The forest is majorly made of individual plants that are the essential component of the ecosystem structure. Individually, every plant has a little impact on the climate, habitat provision and nutrient cycling, and may even be incapable to reproduce.

Although, when millions of plants are brought together, various kinds of ecosystem functions evolve (In the Amazon or Congo basin). The forest canopy filters out around 98% of the sunlight at the ground level, intensely decreasing the daytime temperatures.

It traps insulates and air, raising night temperatures under the canopy and sustaining constant as well as high humidity. Trees usually absorb the energy of tropical storms; ventilate the soil to permit absorption of water and the slow water flows- all of which avert nutrients and soil from being vanished from the system.

Trees usually create the habitat and microclimate important to the soil fauna that aid in recycling the nutrients, enabling their reabsorption by the system. The water reserved by the structure of forest on a regional scale is absorbed and then returned to the atmosphere by evapotranspiration, growing humidity over the forest. More humidity raises the frequency of rainstorms.

Estimations for the Amazon forest recommend that it produce up to 50% of its own rainfall, making the water-dependent species able to thrive there. Devoid of the raised absorption capacity of the soil as well as the evapotranspiration it enables, rainfall would drain into the rivers and be flushed from the system ever.

On a larger scale, forests absorb around 90% of the solar energy that usually strikes their canopies. A huge amount of this is released by the process of evapotranspiration and carried up high in the atmosphere, where it is taken into the zones of temperature, aiding in stabilizing the global climate.

2.4. BIOTIC RESOURCES AND ASSOCIATED ISSUES

There are three major issues related to the biotic resources, as depicted in Figure 2.4.

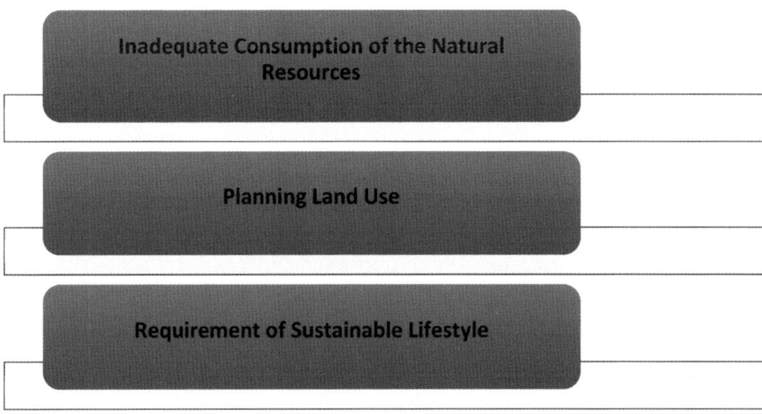

Figure 2.4. *List of major issues related to biotic resources.*

2.4.1. The Uneven or Inadequate Consumption of the Natural Resources

A key part of the biotic resources is consumed in the developed or technologically advanced world now a day. The developing countries such as China and India, overuse various resources. The reason behind this huge consumption is their great human population.

Although, the resource consumption per capita of the developed nations is up to 50 times greater as compared to the case in developing nations. Advanced or progressive nations produce more than 75% of the universal industrial waste as well as greenhouse gases (GHG). Energy obtained from fossil fuels is consumed in greater quantities in the developed nations.

Their per capita food consumption is also huge as well as the waste of a huge amount of food and various other products like packaging material, utilized in the industry of food. For example, the USA with only 4% of the total population of the world, consumes around 25% of the resources present around the globe.

More and is required for producing animal food for the purpose of human consumption instead of growing more crops. Therefore, nations that are majorly dependent upon the non-vegetarian diet require bigger areas for the pastureland as compared to those where people are usually vegetarians.

2.4.2. Planning Land Use

Land in itself is a big resource. It is mainly required for the production of food, animal husbandry, various industries and for improving human settlements. These kinds of rigorous land use are often extended at the cost of wild lands, our enduring forests, wetlands, grasslands, and deserts.

Therefore, it is important to evolve or develop a rational policy of land use that evaluates how much land should be made available for numerous purposes and where it should be located. For example, there are generally alternate sites at which dams or industrial complexes can be constructed, but a natural wilderness cannot be artificially recreated.

Today, scientists and researchers believe that at least 10% of the water and land of each ecosystem should be kept as wilderness for the long-term requirements of securing nature as well as natural resources. As a resource, the land is now under severe pressure because of an increasing land hunger so as to produce enough quantities of food for blowing up the human population.

In addition, it is affected by degradation because of misuse. Water and land resources are usually polluted by industrial waste as well as urban and rural sewage. Increasingly, they are being averted for short-term economic profits to industry and agriculture.

Natural wetlands of the huge value are being exhausted for the agriculture and numerous other purposes. Semi-arid land is being overused. The most destructing change in the land use is illustrated by the speed with which these forests have damaged during past few years around the globe. A forest provides a huge variety of significant services. These consists of numerous processes such as removal of carbon dioxide, maintaining the level of oxygen in the atmosphere, slowing down the erosion, control over the water regimes, and produce distinct products such as fuel, food, timber, medicinal plants, fodder, etc. the loss of these in long term is far greater as compared to the short term gains produced by transforming land to some other uses.

2.4.3. The Requirement of Sustainable Lifestyle

The need for sustainable lifestyles is very important in context of a biotic resource. The quality of human life, as well as ecosystem present on earth, is the major indicators of the sustainable utilization of significant resources. There are strong indicators of the sustainable lifestyle in the human life, such as:

- Improved longevity;
- A rise in knowledge;
- An improvement of income.

These three factors together are known as the 'Human development index (HDI).' The quality of the ecosystems has such kind of indicators that are more difficult to evaluate.

- A stabilized population;
- The long-term preservation of biodiversity;
- The cautious and long-term use of natural resources;
- The inhibition of degradation as well as pollution of the environment.

2.5. EQUITABLE USE OF RESOURCES FOR SUSTAINABLE LIFESTYLES

Decrease in the unequal and unsustainable use of resources and control of the population growth are required for the existence of the country and certainly of the human kind everywhere.

The environment provides with a huge variety of goods and services mandatory for our day-to-day lives, but the water, climate, solar energy and soil, which form the abiotic support that is derived from the nature, are in themselves evenly distributed all around the world.

A modern economic order at the national, as well as global level, must be based on the capability to distribute profits of the natural resources by sharing them equally amongst the nations and communities within various countries.

It is at the local level where people exist by the sale of locally collected resources that the difference is huge. Development has not yet reached them, and they are generally accused of degrading or exploiting natural resources. They should be sufficiently remunerated for the removal of the sources to reserved regions and therefore, develop a huge stake in securing or preserving the natural resources.

There are numerous principles that everyone can easily adopt to bring sustainable lifestyle. This mainly comes from taking care of our earth in every single aspect. A love and respect for the nature is the great sentiment that aids in bringing about a feeling for seeing at how people utilize natural resources in a modern and more sensitive way.

Just think of the beauty of a natural forest and its magnificence, the so clean water of a lake that provide support to life, the area of green grassland, the magnificent potential of the oceans, the clear water of the hillstream, and support the conservation of the wealth of nature. If people respect this, they cannot commit such acts that will degrade the supporting systems of our life.

2.5.1. Environmental Impacts of Resource Use

The utilization of significant resources can definitely lead to a wide range of environmental impacts, such as:

- **Direct Impacts of Extraction:** The impact on the landscape and nature of opencast mining.
- **Disruption of Material Cycles:** It is caused by the introduction of unavailable matter into the atmosphere, major movements of hazardous materials, and the loss of natural areas as an outcome of erosion and deforestation.

Various other impacts on the environment are linked with the use of biotic resources, such as the use of pesticide for the production of food, and the acidification is mainly caused by the fossil fuel combustion.

The extraction of minerals and fossil fuels causes large impacts on the environment, and it is linked with the regular threats that have distant impacts. The major example of this is the huge disaster of Spain in the year 1998 and Romania/Hungary in the year 1999 and the oil spills from Russian oilfields in the year 2000.

Accidents continue to take place during the distribution and transportation of oil, and they have various impacts on the marine environment. The utilization of biotic resources is the most crucial human activity in context of global environmental effects.

2.5.2. Ecological Limits

The environmental impacts of utilization reserves of the biotic resources are so huge that numerous ecological limits are easily surpassed.

- **Greenhouse Effect:** As per the Intergovernmental Panel on Climate Change (IPPC), the sharp greenhouse effect must be decreased by 50–70% from the year 1990 levels to retain the effects of climate change in the manageable proportions.

Signing the Kyoto protocol in the year 1997 was the first step for the industrial nations. Until then, very less attempt has been made to attain a decrease of 50%.

- **Biodiversity:** At the 'Earth Summit' held in Rio de Janeiro (UNCED) in year 1992, it was established that the current levels of biodiversity all around the world must be conserved as far as possible. Every country is equally responsible for the preservation of its own biological diversity. On the other hand, the loss of biodiversity is still taking place in spite of a huge number of successes.

2.5.3. The Challenge Facing the World

The world is facing two major challenges regarding the use of resources. First is that of attaining a fair distribution of the resource utilization. Attaining this will need an exact 'dematerialization' of consumption. Around 70% of the greenhouse effect is caused by the overuse or misuse of fossil resources.

Out of this 70%, more than half is utilized in the production process of raw materials, components, and various products in different industries. The strength of decreasing the greenhouse effect by changing the system of energy is restricted by the demand of effectiveness of material in the process of production of goods.

To attain a 50 to 75% decrease in the emission or production of GHG will be very tough without a huge increase in proficiency of the material system. The fundamental position of Dutch policy on the greenhouse effect is that these issues or problems should not be transmitted to food production and biodiversity.

Dematerialization of the use of foodstuffs as well as wood will help the barren land for the biofuel crops and storage of carbon required to fight global warming. This must not further compete for land with the increasing demand in developing nations for good quality food and timber.

The second major challenge, which the world is currently facing, is to produce such raw materials that are in the accepted in the local environment as well as social conditions and it must contribute to attain the targets of global sustainability.

For the extraction of minerals and various fossil resources, the local environmental impacts should be decreased to the minimum level, and the international conventions must be respected. The production of food and timber must consider the targets of global sustainability, such as deductions in the global warming and the hindrance of soil erosion, clear felling, and biodiversity loss.

2.6. CONSUMPTION OF RESOURCES

Nowadays, the consumption of resources has increased a lot. The main reason behind this great consumption is the rapid increase in the population. Consumption is directly proportional to the number of populations. People are using the most significant resources in such a way that they have started getting depleted.

There are major differences in per capita consumption of biotic resources between different nations and regions of the world. People living in rich nations usually consume up to ten times more as compared to those who are living in the poor nations.

The most resource-intensive areas of life in Europe are food, housing, and transport. These three areas make up almost 70% of our total resource consumption. Universal resource consumption equals global resource extraction: the world economy uses almost 60 billion tonnes of resources every year to produce the goods and services, which people consume.

Huge differences in per capita consumption: In Europe, around 36 kg of the biotic resources are extracted per day per person, exclusive of the unused resource extraction, whereas total of 43 kg are consumed per day per person. Therefore, Europeans require resources imported from other regions of the world to maintain the level of consumption.

Consumption is even much greater in various other regions of the world. An average North American consumes almost 90 kg per day whereas inhabitants in Oceania around 100 kg per day. On average, in comparison to Europe, people living in these continents have big houses, consume more amount of meat and drive bigger cars. These lifestyle differences increase the consumption of resource. In various other regions, very less amount of resources is consumed. In Asia, the consumption of resources is almost equal to the extraction of resources, which is around 14 kg per day per person. The average consumption of resources in case of Africa is just 10 kg per person per day as compared to its extraction being 15 kg per person per day.

It exactly means that in Europe, people consume three times of the resources as an inhabitant of Asia and four times of the resources as an average African. Inhabitants of various other nations consume around ten times more as compared to people living in developing nations.

The difference would be quite high if unused resource extraction were comprised in these calculations. Almost one-third of the biotic resource consumption is caused by the infrastructure and housing. Biotic resources are mainly required for the biosphere. Additionally, people need petroleum and gas for the purpose of cooking, transportation, etc.

Drinking and eating usually accounts for almost one-quarter of the resource utilization. It generally consists of food and beverages bought in shops and consumed in various restaurants and hotels.

The industry of food and drink needs various resources like energy, machinery, agricultural products, etc. to produce those products that people purchase from the shops. In addition to that, the network of retail uses a lot of transport as well as cooling.

All of these resources use is the major component of the consumption category. Plants, animals, bacteria, fungi, etc. makes the significant elements of the biotic resources.

2.7. GLOBAL PERSPECTIVE OF NATURAL RESOURCES

The intense or powerful negotiations of the year 2015 led people to question the models of development, given the outcomes of the deepening of inequalities as well as climate change. Two chief concerns mainly contribute to dominate the agenda that are:

- How to fulfill the need or demand of rapidly increasing population? and
- How to do so while preserving the natural resources and environment for the upcoming future generations?

People currently living on the planet are very much hungry for the land and thirsty for the water, and strains proliferated over the resources required to meet their requirements in context of food, heating, housing, and environment.

All of these activities that are increasing competition amongst each other at the local level have a major impact on the state of natural resources, at sometimes flaring up into the crises as well as violent conflicts, which impend the security and peace of the regions or countries.

If the present trend for waste and putting pressure on the ecosystem, land, water, primary forests, living conditions of people could undertake a deep transformation in the coming future, with profound outcomes.

Thus, a paradigm shift is essential as well as desirable, massive, and critical efforts will be required for the combined management of the natural resources at local as well as global level. By no means, the Mediterranean is exempt from this alarming state of affairs.

In contrast, it reflects all kind of tensions that rotate around the management of biotic resources and agriculture. However, the demographic weight is decreasing universally; the increasing population of the world continues to employ a great pressure on already rare biotic resources.

To be sure, the population growth is coupled with the strong coastal urbanization, which is leading to the overuse or misuse of the resources and is conceding the potential for development in the specific region.

2.7.1. Some Global Perspectives on the State of Natural Resources

The planet is being majorly subjected to the variety of transitions or alterations all at the similar time. Two of them have a severe impact on the natural resources. These two are demographic transition and food transition.

Demographic transition is taking the population of world to new and high levels, whereas the food transition is generating unique production requirements with an increase in the daily consumption as an outcome.

Therefore, people currently living on the continents have increased their utilization of water and land and enlarged the land area under cultivation. Once in a while, this has been at the cost of terrestrial biodiversity and forests, which is disappearing in an alarming way in few parts of the world.

In an effort to meet the huge demands of natural resources, people are moving to the seabed of the planet, which provides great reserves of food, and biodiversity that is going to be affected in some of the regions. The demographic transition linked with the global increase in the average income tens to place a huge burden on the energy resources.

2.7.1.1. Marine and Fisheries Resources

For a substantial share of the population of the world, fisheries, and marine resources are key to prosperity and subsistence. Their misuse or overuse from the shores of the Mediterranean to the inland seas of Europe and Asia and the continental waters of Brazil, Canada, and China, has facilitated great civilizations and major powers to flourish.

The contribution of aquaculture and fisheries products to the human diets has been documented by the member states of the Committee on World Food Security, proof of the major significance of this sector in contending hunger as well as malnutrition.

Although, because of human activities, resources of fisheries have become delicate and are generally overused to satisfy the requirements of an increasing population, whose members have never before consumed huge verities of food products drawn from rivers and seas.

Technological advancement, increasing density of population alongside the coasts and rising urbanization all enhance the pressure on resources as well as the biological diversity of the marine environments.

Dramatically affected by the overuse and pollution, these regions are being distorted by a broad range of the economic activities like extraction of minerals, sand, gas, and oil, fishing, as well as being susceptible by the climate change, whose adverse effects are now making itself felt on the ecosystem and aquatic resources.

According to the report of the Food and Agricultural Organization, the state of resources and fisheries stock gives significant cause for concern. Though global production of the marine fisheries grasped its peak in the year 1996 at 86.4 million tonnes, total fisheries production, which increased around 93.7 million tonnes in the year 2011, is still increasing.

According to the FAO State of World Fisheries and Aquaculture report published in 2014, the Northwest Pacific revealed the highest level of production in the year 2011, followed by the Southeast Pacific. In the Indian Ocean, captures remain to increase, with a growth of 17% between the years of 2007 and 2011. The fishery resources in the Atlantic have greatly suffered from high levels of misuse.

Though some stocks in the North Atlantic region have depicted signs of recovery because of enhanced management systems, in the southeast, 55% of stocks examined were being fished at a level that is not naturally viable in the long term.

At the universal level, overexploited or misused stocks have increased since the time of the 1970s, accounting for total of 28.8% of the harvested fish stocks in the year 2011. In the Mediterranean and the Black Sea, 52% of stocks graphed were fished at indefensible levels.

For instance, those of cod and mullet are damaged, while those of sole, anchovies (pelagic species) and sardines are considered completely misused. According to the Red List of threatened species drawn up by the International Union for the Conservation of Nature (IUCN), 43 species of native marine fish are susceptible at the regional level.

The position of Bluefin tuna in the Mediterranean, whose reproductive power has decreased by 50% over the past four decades, remains tremendously worrying, in spite of the 2006 launch of a recovery plan, with revised fishing quotas, which has led to a small progress.

2.8. BIODIVERSITY: TOWARDS A "SIXTH MASS EXTINCTION"

The assessment of the millennium ecosystem has revealed the degree of outcomes of modifications to the ecosystem to meet the requirements of the total population of the world. As per the final report, the transformation of ecosystem during the past five decades has taken place rapidly in the history of humanity.

These alterations have made it possible to meet the requirements of water, food, energy, fiber, and rubber, but have led to irreversible as well as substantial losses of the aquatic and land-based life.

The protection or conservation of biodiversity as defined in the Convention on Biological Diversity 1992 as "the variability among living organisms from all sources including, inter alia, terrestrial, marine, and other aquatic ecosystems and the ecological complexes of which they are a part; this includes diversity within species, between species and of ecosystems," has been retained on the schedule of the international community. Every day, science is making improvement in characterizing, discovering, and evaluating endangered species. There is a significant scientific controversy regarding the speed of the extinction and disappearance of the living species.

However, whether the rate is 50, 100, or 1,000 times more than that at the start of life on the earth, the significances for the human race as well as its environment are disastrous. Various scientists are openly talking about the sixth mass extinction, the last one being during the Cretaceous age, which saw the vanishing of the dinosaurs, almost 65 million years ago. Amongst many ecosystems that are exposed are coral reefs.

They have great levels of the biodiversity as compared to the tropical forests. According to a researcher of Australia, 30% of these coral reefs have been severely destructed by disease, pollution, and fishing.

Around 35% of the mangroves have also been vanished during the last twenty years because of the storm, overuse or misuse and transformation into aquaculture. The current version of the IUCN Red List, drawn up in the year 2015, categorizes 22,784 species out of 77,340 studied as "exposed with extinction": 41% are amphibians, 31% sharks, 25% mammals, and 13% birds.

Megafauna (elephants, polar bears, rhinoceroses, etc.) and invertebrates (butterflies, ladybirds, spiders, etc.), which have dropped by 45% since the year 1980, have perceived the sharpest drop. There are five main sources of pressure weighing on the biodiversity:

- Degradation of natural environments (e.g., deforestation);
- Overexploitation or misuse of the natural resources (e.g., fisheries resources);
- Introduction of invasive or hostile species (farmed fish, exotic pets, shellfish introduced into the Mediterranean through the Suez Canal and ballast water, etc.);
- Pollution (caused by heavy metals); and
- Climate change.
- Because of the opening up of the commerce, improved trade along with the great movement of goods as well as people in the Mediterranean, the threats of presenting and distributing dangerous organisms is continuously increasing at an alarming rate.

The battle to stop the losses or harms associated with the dangerous organisms is specifically important for the conservation of food security in the Mediterranean, which is a net trader of cereals. In context production, the agricultural intensification or growth seen in few of the nations in the past few decades have severe influences on the diversity of genetic resources for food as well as agriculture.

Remarkably, amongst various examples, a major example is the selection of few breeds of dairy cow, that are causing other evidently less productive breeds to be abandoned as well as disappear, and a trend for few regions to concentrate on producing a restricted number of crops.

2.9. GLOBAL CHALLENGE TO FEED THE WORLD AND MAINTAIN THE NATURAL RESOURCE

Erosion of the soil has an adverse impact on the production of food, mainly considering that the landmasses produce more than 99% of the total food of the world, while less than 1% of food is acquired from the aquatic habitats and oceans.

The Global Land Assessment of Degradation approximates that around 2 billion hectares have been degraded or ruined since mid-century all around the world, that is, 22% of all pasture, cropland, forest, and woodland. The data for cropland alone is even poorer.

There are around 1.5 billion hectares globally in the cropland, of which around 38% are damaged to some extent. The cumulative productivity loss for the cropland from the phenomenon if soil degradation over the past five decades is estimated to be around 13%. According to WHO, out of the 6 billion people living on earth, 3 billion are malnourished or underweight.

According to the report of FAO, the global production of food will be required to be increased by 60% to remove the nutrition gaps, to cope with a rapid growth in the population size and to accommodate huge changes in diets over the next thirty years.

The FAO estimates that a huge number of farmers are considering conservation farming and it is undertaken on around 58 billion hectares of land. The process of conservation farming decreases the water loss, chemical inputs, fuel consumption, and soil erosion increases the production and protects the structure and biology of the soil (Figure 2.5).

Figure 2.5: Conservation farming.

Source: https://commons.wikimedia.org/wiki/File:Conservation_farming_02.jpg

These farming practices include numerous methods that do not have any adverse effect on water and land resources. For instance, in Africa, Latin America, and Asia, diversification into the perennial crops is protecting soils; in Jordan, Syria, Argentina, Southern Africa and Mexico, the schemes of dryland range rehabilitation are showing positive results.

2.10. ROLE OF AN INDIVIDUAL IN CONSERVATION OF NATURAL RESOURCES

Until now, people acted as if they could go forever on exploiting the ecosystem and biotic resources such as water, soil, and forests present on the surface of earth, and obtaining the fossil fuels and minerals from underground.

In the past few decades, it has increasingly become evident that the universal ecosystem has the ability to sustain or maintain an inadequate level of utilization. Biological systems cannot go on replacing resources if they are misused or overused.

As a crucial point, growing pressure disrupts their natural balance. Traditionally, the biological resources are classified as renewable- such as those from the forests, oceans, wetlands, and grasslands are being destructed by the misuse or overuse and may be destroyed permanently.

None of the natural resource is limitless. "Non-renewable" resources will be exhausted rapidly if people continue to utilize them as intensively as at present. The two most destructing factors leading to the present depletion of every form of natural resources are raising consumerism on the part of the rich sections of society, and fast population growth.

Both of the factors are the consequence of choices we make as individuals. As a responsible individual, people need to decide:

- What will we leave for our children?
- Is there any kind of material gain someone else's loss?

Figure 2.6: Role of people in conservation of resources.

Source: https://www.wpafb.af.mil/News/Article-Display/Article/1545053/wetlands-and-prairies-natural-resources-division-is-committed-to-conservation/

The greed for material goods has really become a way of life for huge number of people in the developed world. Population growth, as well as the subsequent shortage of the resources, strictly affects people living in the developing nations.

In various countries that are rapidly developing as well as suffering from the explosion of population, both are major factors that are responsible for the degradation or destruction of environment. People must ask themselves if they have reached a crucial flashpoint perhaps, at which point economic development affects people lives more adversely as compared to the advantages it provides (Figure 2.6).

2.10.1. Conservation of Biodiversity and Wildlife

The term "conservation" is uncertain or indefinite. For different people, it has a different meaning. The ambiguity or uncertainty of the term conservation is derived from its two roots- one in the management of resource and another in the natural history.

The major concern with the management of resource, which considered that the biological resources had to be sustained in such a way that they do not get wasted and assured that they did not become extinct, is a long-standing one and was clearly said by the North American forester Gifford Pinchot who compared conservation with the systematic exploitation or misuse of natural resources.

The natural history roots of the conservation are very strong and generally expressed by the concern at the degradation of species as well as wilderness aspect of the natural landscapes. It is this general perception of conservation that is the major source of the public support for the action of conservation.

The development of the biology conservation concept is strictly interrelated with that of the biodiversity. Similar to the conservation biology, the concept of biological diversity later contracted as biodiversity. It was developed in the phase of 1980s though its origins go back much earlier.

Again, similar to the conservation biology, biodiversity is a synthetic discipline. It serves as a focus where numerous disciplines, as well as activities, meet and interact. Certainly, the disciplines serving the biodiversity are generally the same- taxonomy, genetics, ecology, and biology.

The concept of biodiversity has increased into its prominence in an unusually short period. Perhaps, it is the dominant concept in the environmental planning, thinking, and action, and the entry and signing into force of the Convention on Biological Diversity has placed it on the international as well as political stages.

There is still a considerable debate as to what is biodiversity, whether it is a meaningful notion or just a passing fashion, whether it can be regarded as a severe discipline and whether its assets all the attention that it appears to attract.

The basic issue is the striking simplicity of the concept: biodiversity as the 'variety of life' is the type of general instinctive definition that most of the people share. It is when people try to apply rigorous and precise definitions, and then people get a great variety of different outlooks regarding the biodiversity.

Numerous researchers recommend that these can be widely grouped under three major headings: those that consider biodiversity as a concept; those that consider it as a measurable entity; and those that consider it as a social or political construct.

The most recently employed definitions differentiate between the major levels or components of the biodiversity that can be recognized- genes, species, ecosystems-, and these are further recognized in the definition given in the Article 2 of the Convention on Biological Diversity, as follows:

"Biological diversity means the variability among living organisms, from all sources, including, inter alia, terrestrial, marine, and other aquatic ecosystems and the ecological complexes of which they are part; this includes diversity within species, between species and of ecosystems."

Wildlife conservation is defined as an activity in which humans usually make cognizant efforts to protect or preserve plants and animal species. It is really very important to conserve biodiversity as it plays a significant role in sustaining the ecological balance and contribute to the quality of human life.

The phrase "wildlife conservation" conceptually increases a valuation process in which it is certain that something, in this case, must be observed. The decision of conserving the wildlife mainly requires valuation and justification of that which is to be protected or conserved.

Justifications for the conservation process can be broadly thought as falling into two major categories:

- The first category presumes that there are possibly detectible advantages to be derived by conservation; and
- The second category is completely based on the concept that

organisms have a right to survive as they have already survived for a long period, so that it is quite tough to define identifiable profits to be derived by the existence of these organisms.

In terms of wildlife conservation, valuation must identify differences between held and assigned values of the wildlife, and should always consider what kind of values cause attitude towards the wildlife and what kind of wildlife and their number, in what settings in which we find them, and what opportunities they provide.

Numerous researchers traced the growth and development of the wildlife conservation and management by reviewing different textbooks and other significant documents starting with the work of Aldo Leopold and leading up into the phase of 1980s.

Scholars contended that the most significant notion for the contemporary wildlife management as well as conservation was that wildlife management and conservation involves "the art and science of making decisions as well as taking actions to influence the structures, dynamics, and relations of habitats, populations, and people to attain specific human goals by means of the wildlife resource."

2.11. THE FUTURE OF NATURAL RESOURCES: IMPORTANT, CONTESTED, UNCERTAIN, AND COMPLEX

There is a major concern regarding the future availability of biotic resources, which has increased severely in the early 21st century. These resources actually comprise water, food, energy, and minerals. At the level of material, this was mainly because of the rapid increase in the demand, specifically in Asia, and subsequent price rises that wiped out the deductions in the average price of a commodity from past ten decades.

These concerns have been intensified by the major issues of scarcity, which was once observed as a temporary or geographically contained phenomenon, which have become universal in scope and impact long-term views. Together, the threat of attaining a point of no return in the environmental deprivation has not just evolved concerns that a degraded ecosystem could impend the availability of biotic resources, but it has also given rise to various concerns that recent abundance in few stocks could damage the long term sustainability of the environment (Figure 2.7).

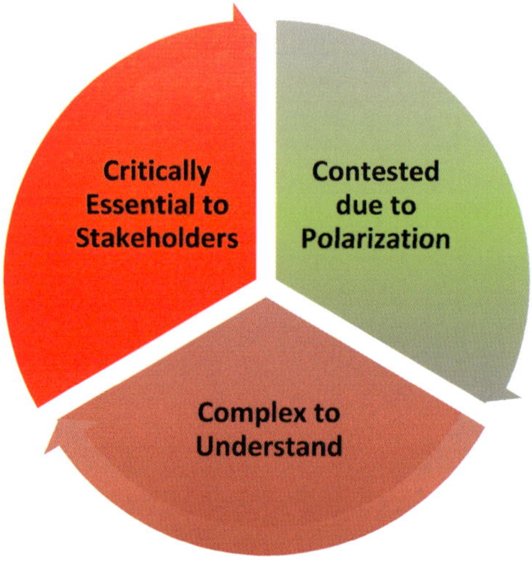

Figure 2.7: Subsequent image of the future availability of significant resources.

2.11.1. Critically Essential to Stakeholders

In spite of the exhortations by few to decouple the economic growth from the growing utilization of various resources, the current economic system of the world requires easy access to the essential biotic resources so as to provide population with a minimal quality of the life.

Scarcities in the crucial input could cause effective damage to the social as well as economic systems, leading to the geopolitical conflict, social unrest, and political instability.

Such kind of efforts are distributed unevenly; as with the universal food markets in past few years, volatile, and high prices of commodity can generate existential concerns for the low-income families, donate to the political instability and the influence trends like global migration.

2.11.2. Contested Due to Polarization

Over the first ten years of this century, an outbreak of the academic papers, investors advice, statements from the civil society, alarming reports from the media have resulted in a rising polarization of the outlooks on the matter, with the perspectives of stakeholders harshly divided as either bullish or bearish on the future availability of resource.

The outcome has been a scrambled image favoring antipathy and reasons associated to the personal opinion, rather than a mutual understanding and the cooperative solutions to the resource challenge of the world.

2.11.3. Complex to Understand

Availability of the natural resource is the significant function of the supply as well as demand of the resources that are exposed, created, processed, transported, and consumed in the complicated value chains, an essential portion of which is global.

Though these value chains appear to work as markets for various consumers, they all usually suffer from the alterations at distinct points, recognition to the monopolistic structures, controlled supply routes and the government intervention such as taxation and subsidies.

Additionally, as around all of the natural resource value chains are based on the physical interference from climate, weather, and the political uncertainty, pricing on the universal markets is very sensitive to the actions of investors as well as traders unresponsive in the physical delivery, and hence exposed to the usual views on the universal economic growth.

This major complexity usually leads to the substantial uncertainty on the future demand as well as supply of numerous resources, that are based on the sensitivity of consumer price and the possible issues related to the process of supply.

On the upside, surprises are huge in number, as abundant rainfall, new discoveries, and technological breakthroughs can rapidly cause a shift to abundance from scarcity. Almost every prediction or assumption of the availability of resources and prices use the current past as a key guide, both of them are highly volatile and have verified as impossible to forecast reliably over the long-term period.

REFERENCES

1. *Biotic Resources*, [eBook] (p. 18). Retrieved from: http://www.uvm.edu/~jdericks/EE/ch6.pdf (Accessed on 21 May 2019).
2. Brun, M., Blanc, P., & Otto, H., (2016). *Global Perspective of Natural Resources* [eBook] (p. 17). HAL. Retrieved from: https://hal.archives-ouvertes.fr/hal-01620873/document (Accessed on 21 May 2019).
3. *Conservation of Natural Resources Including Forestry and Wildlife*, (2019). [eBook] (p. 30). Retrieved from: http://www.moef.nic.in/report/0304/chap-03.pdf (Accessed on 21 May 2019).
4. Heywood, V., & Iriondo, J., (2003). *Plant Conservation: Old Problems, New Perspectives*. Retrieved from: https://www.researchgate.net/publication/222699533_Plant_conservation_Old_problems_new_perspectives (Accessed on 21 May 2019).
5. *Natural Resources*, (2004). [eBook] (p. 37). Retrieved from: https://www.ugc.ac.in/oldpdf/modelcurriculum/Chapter2.pdf (Accessed on 21 May 2019).
6. Tidball, K., (2014). Wildlife conservation. *Encyclopedia of Quality of Life and Well-Being Research*, 7130–7131. doi: 10.1007/978-94-007-0753-5_3241.
7. *What are Biotic Resources?* (2019). Retrieved from: https://www.thebigger.com/biology/resources/what-are-biotic-resources/ (Accessed on 21 May 2019).
8. Williams, J., Bordas, V., & Gascoigne, H., (2004). *Conserving Land and Water for Society: Global Challenges and Perspectives* [eBook] (p. 17). Retrieved from: http://www.tucson.ars.ag.gov/isco/isco13/PAPERS%20R-Z/WILLIAMS.pdf (Accessed on 21 May 2019).
9. World economic forum, (2014). *Resource Scarcity the Future Availability of Natural Resources a New Paradigm for Global Resource Availability* [eBook] (p. 56). Retrieved from: http://www3.weforum.org/docs/WEF_FutureAvailabilityNaturalResources_Report_2014.pdf (Accessed on 21 May 2019).

3

Abiotic Resources and Their Global Challenges

CONTENTS

The world is developing at a rapid pace and to keep itself functioning, and it needs to have a continuous supply of energy. The energy needs to be generated from the sources that are available to the mankind. These sources can both be renewable and non-renewable. The world now looks forward to the renewable energy resources, as it realizes that the non-renewable ones are depleting at a very fast pace, and to fulfill the energy demands of the whole world, these resources are nowhere near to being enough. The renewable forms of energy constitute resources in both biotic and abiotic forms. The biotic forms of energy have been deeply dwelled upon in the previous chapter. Whereas, this chapter takes the readers through the possibilities that lie across the globe when abiotic components of the ecosystem build up to supply the energy demands in some way or the other. This chapter also makes the readers aware of the challenges that are posed when these sources are used to generate and harness energy.

3.1. INTRODUCTION

Nature gives us a variety of products and services essential for the everyday lives. These natural resources incorporate, air, water, soil, minerals, alongside the atmosphere and solar energy, which form the non-living or 'abiotic' components of nature.

The 'biotic' or living components of nature comprise of plants and animals, including microorganisms. Plants and animals can just survive as networks of various organisms, all closely connected to each other in their own living space, and requiring particular abiotic environments.

In this way, forests, fields, deserts, mountains, streams, lakes, and the marine environment all makeup habitats for specific networks of plants and animals to live in. Collaborations between the abiotic parts of nature and some particular living organisms together form biological systems of different kinds. Huge numbers of these living organisms are utilized as the food resources.

Others are connected to the food in a lesser direct manner, for example, pollinators, and dispersers of plants, soil organisms like worms, which reuse supplements for plant development, and parasites and termites that break down dead plant material with the goal that smaller scale organisms can follow up on the debris to transform soil supplements.

Around 10,000 years before, when humanity transformed from a hunter collector, living in wild regions, for example, forests, and meadows, into an agriculturist and pastoralist, humans started to transform the earth to suit the possess necessities.

As the capacity to develop food and utilize domestic animals developed, these 'characteristic' biological systems were formed into agricultural land. Most conventional agriculturists relied widely on rain, streams, and rivers for water. Later they started to utilize wells to tap underground water sources and to store water and made inundated land by setting up dams. As of late humans started to utilize manures and pesticides to additionally support the production of food from a similar expanse of land.

Anyway, humans presently understand that this has prompted a few unwanted changes in the earth. Humanity has been abusing and draining normal resources. The over-serious utilization of land has been found to deplete the ability of the environment to help the developing requests of an ever-increasing number of individuals, all requiring progressively escalated utilization of resources. Modern development, urbanization, populace development, and the gigantic increment in the utilization of purchaser products have all put further weights on the earth. They make extreme amounts of strong waste. Contamination of air, water, and soil has started to genuinely influence human wellbeing.

3.2. SOLAR ENERGY

Solar energy is the energy gotten directly from the sun, and it is an amazingly clean and inexhaustible type of energy. The sun is the most plenteous source of energy on Earth (Figure 3.1).

Figure 3.1: Solar collectors at NASA.

Source: https://imagecache.jpl.nasa.gov/images/edu/activities/solarenergy-640x350.jpg

The primary employment of solar energy are: water heating (solar heat collectors convert the sun's beams into heat), generation of power (the photovoltaic (PV) cell changes over sunlight directly into power), heating buildings (solar heat collectors convert the sun's beams into heated air) and desalination of seawater.

Solar panels convert solar energy into DC (locale flow) power, which enters an inverter. The inverter transforms DC power into AC (substituting flow) power required by home apparatuses or lights.

At the point when more solar energy is delivered than the amount required, it very well may be put away in a battery as DC power. One noteworthy favorable position of solar energy is that it is accessible to everybody and can be harnessed by people almost everywhere, subsequently, making electricity dissemination over a huge amount of area, pointless.

3.2.1. What Is Solar Energy?

Consistently, the sun transmits (provides) a colossal amount of energy called solar energy. It emanates more energy in one day than the world uses in a single year. This energy originates from inside the sun itself.

Like most stars, the sun is a major gas ball made up for the most part of hydrogen and helium gas. The sun makes energy in its internal center in a procedure called atomic combination. It takes the sun's energy only barely eight minutes to venture to every part of the 93 million miles away situated Earth.

Solar energy goes at the speed of light, or 186,000 miles for every second, or 3.0×10^8 meters per second. Just a small part of the evident and visible energy in the form of light, that the sun emanates into space ever reaches the Earth, yet that is all that could possibly be needed to supply all the energy needs.

Consistently enough solar energy reaches the Earth to supply the world's energy requirements for a year. Solar energy is viewed as a sustainable power source because of this reality. Currently, individuals utilize solar energy to heat buildings and water and to create power. Solar energy represents a small amount of U.S. energy deposits – even lesser than 1%. Solar energy is for the most part utilized by residential establishments and to produce power.

3.2.2. Solar Collectors

Heating with solar energy isn't as simple as one may might suspect. Catching sunlight and utilizing it in some form, is troublesome in light of the fact that the solar energy that reaches the Earth is spread out over a huge region.

The sun does not provide that much energy to one particular spot at any given time. The amount of solar energy a zone gets relies upon the time of day, the period of the year, the clouds in the sky, and the distance one is near to the Earth's Equator.

A solar collector is one method to catch sunlight and transform it into usable heat energy. An enclosed vehicle on a bright and hot day resembles a solar collector. As sunlight goes through the vehicle's windows, it is consumed by the seat covers, walls, and floor of the vehicle.

The consumed light changes into heat. The vehicle's windows let light in; however, they don't give all the heat a chance to get out. An enclosed vehicle can get extremely hot.

3.2.3. Solar Space Heating

Space heating stands for heating the space inside a building. Currently, numerous homes utilize solar energy for space heating. An inactive solar home is intended to let in as much sunlight as could reasonably be expected. It resembles a major solar collector.

Sunlight goes through the windows and warms the walls and floor inside the house. The light can get in; however, the heat is caught inside. A passive solar home does not rely upon mechanical machines, for example, siphons, and blowers, to heat the house, while dynamic solar homes do.

3.2.4. Solar Water Heating

Solar energy can be utilized to heat water. Heating water for washing, dishwashing, and garments washing is the second biggest home energy cost. Introducing a solar water radiator can lessen one's water heating bill by as much as 50%. A solar water heater works a great deal like solar space heating.

In some nations, a solar collector is mounted on the south side of a rooftop where it can catch sunlight. The sunlight warms water in a tank. The high-temperature water is channeled to fixtures all through a house, similarly as it would be with a conventional water heater.

3.2.5. Solar Electricity

Solar energy can also be utilized to create power. Two different ways to make power from solar energy are PV and solar heat systems.

3.2.6. Photovoltaic (PV) Electricity

PV originates from the words photo, which means light, and volt, a unit to measure electricity. On some occasions, PV cells are referred to as PV cells or solar cells in short. One is likely to be acquainted with PV cells.

Solar controlled toys, calculators, and roadside phone call boxes all have incorporation of solar cells to transform over sunlight into electricity. Solar cells are comprised of silicon, a similar substance that forms sand.

Silicon is the second most widely found substance on Earth. Solar cells can supply energy to anything that is controlled by batteries for electric power. Power is delivered when light energy from the sun strikes the solar cell, making the electrons move around.

The activity of the electrons begins an electric flow. The transformation of sunlight into electricity happens quietly and in a quick instant. The panel has no mechanical parts that can damage themselves. When compared with different methods for making power, PV systems are costly, and numerous panels are expected to rise to the power created at different kinds of plants.

It can cost 10 to 30 pennies for each kilowatt-hour to deliver power from solar cells. Quite a many people pay their electric companies about 12.6 pennies per kilowatt-hour for the power they use, and the huge companies set up in industrial areas pay lesser for the same amount of power used.

Solar equipment is generally used to create power in remote territories that are far from electric power cables. In 2015, the Desert Sunlight solar project in California opened. It is among the biggest PV plants on the planet, creating 550 megawatts of power—enough to power more than 150,000 homes.

3.2.7. Solar Thermal Electricity

Like solar cells, solar heat systems, additionally known as concentrated solar power (CSP), utilize solar energy to create power, however in an alternate manner. Most solar heat systems utilize a solar collector with a reflected surface to center sunlight onto a recipient that warms a fluid. The super-heated fluid is utilized to make steam to create power similarly that coal plants do. There are CSP plants in California, Arizona, Nevada, Florida,

Colorado, and Hawaii. A component of the world's biggest CSP offices is situated in California. Solar energy has extreme potential for what's to come. Solar energy is free, and its provisions are boundless.

It doesn't contaminate or generally harm nature. It can't be constrained by any one country or industry. In the event that humans can improve the innovation to harness the sun's gigantic power, humans may never confront energy deficiencies again.

3.3. WIND

Wind power is the transformation of wind energy into progressively valuable buildings. Most present-day wind power is created as power by changing over the turn of turbine edges into electrical flow by help of an electrical generator. In windmills (a lot older technology), wind energy is utilized to turn mechanical hardware to do physical work, such as pounding grain or siphoning water.

Interest for wind came about on the grounds that it is an extremely clean type of energy. The oil depletion of the 1970s facilitated interest for wind and other alternate energy sources. In spite of innovative work cuts, there is significant wind research and use currently (Figure 3.2).

Figure 3.2: Wind farm.

Source: https://cdn.pixabay.com/photo/2018/10/23/22/07/wind-3769141_960_720.jpg

Numerous nations have sent wind innovation and use wind hardware to pick up energy. Frequently, power is increased through wind farms – huge collections of wind turbines. Wind power is utilized in large scale wind farms for national electrical matrices just as in small individual turbines for giving power to rural homes or grid-confined areas.

Some individuals have protested wind energy, as they neutralize the greenhouse impact whenever used to supplant the utilization of fossils for fuel. All things considered, some people state that the wind farms are excessively loud and cause traffic clog in the areas near to them. Others complain that windmills kill winged animals.

Wind energy offers numerous benefits, which clarifies why it's one of the quickest developing energy sources on the planet. Research endeavors are done to cater to the challenges for more prominent utilization of wind energy. Further are given the advantages of wind energy and a component of the challenges it is attempting to survive.

3.3.1. Beneficial Aspects of Wind Energy

- Wind power is cost-effective. Land-based utility-scale wind is one of the most low-priced energy sources accessible. Currently, costing somewhere in the range of two and six pennies for each kilowatt-hour, contingent upon the wind resource and the specific venture's financing. Since the power from wind farms is sold at a fixed cost over a significant lot of time (for example 20+ years) and its fuel is free, wind energy mitigates the value vulnerability that fuel costs add to conventional sources of energy.
- Wind develops jobs. The U.S. wind segment utilized more than 100,000 specialists in 2016, and wind turbine professional is one of the quickest developing American employments of the decade. As indicated by the Wind Vision Report, wind can possibly bolster in excess of 600,000 employment opportunities in assembling, establishment, upkeep, and supporting administrations by 2050.
- Wind empowers U.S. industry development and U.S. competitiveness. Wind has a yearly monetary effect of about $20 billion on the U.S. economy. The United States has an immense local resource and a very talented workforce, and can contend worldwide in the clean energy economy.
- It's a perfect fuel source. Wind energy doesn't dirty the air like power plants that depend on ignition of petroleum derivatives,

for example, coal or flammable gas, which radiate particulate issue, nitrogen oxides, and sulfur dioxide—causing human medical issues and financial harms. Wind turbines don't deliver environmental discharges that reason corrosive rain, brown haze, or ozone-harming substances.

- Wind is a local source of energy. The country's wind supply is bottomless and boundless. In the course of recent years, aggregate wind power limit in the United States expanded a normal of 30% every year, and twist currently has the biggest inexhaustible age limit of all renewables in the United States.
- It's feasible. Wind is really a type of solar energy. Winds are brought about by the heating of the air by the sun, the revolution of the Earth, and the Earth's surface inconsistencies. For whatever length of time that the sun sparkles and the wind blows, the energy delivered can be harnessed to send power over the matrix.
- Wind turbines can be based on existing homesteads or farms. This incredibly benefits the economy in provincial territories, where a large component of the best wind locales is found. Ranchers and farmers can keep on working the land in light of the fact that the wind turbines utilize just a small amount of the land. Wind power plant proprietors make lease installments to the rancher or farmer for the utilization of the land, giving landowners extra pay.

3.3.2. Challenges of Wind Energy

- Wind power should, in any case, contend with conventional generation sources on a cost premise. Contingent upon how vivacious a wind site is, the wind homestead probably won't be cost competitive. Despite the fact that the expense of wind power has diminished drastically in the previous 10 years, the innovation requires a higher beginning venture than fossil-filled generators.
- Good wind destinations are frequently situated in remote areas, a long way from urban communities where the power is required. Transmission lines must be worked to bring the power from the wind homestead to the city. However, assembling only a couple of as of now proposed transmission lines could altogether decrease the expenses of growing wind energy.
- Wind resource advancement probably won't be the most productive utilization of the land. Land appropriate for wind-

turbine establishment must contend with alternate uses for the land, which may be more exceedingly esteemed than power age.

- Turbines may cause commotion and aesthetic contamination. Despite the fact that wind power plants have generally small effect on the earth contrasted with ordinary power plants, concern exists over the commotion delivered by the turbine edges and visual effects to the scene.
- Turbine sharp edges could harm neighborhood untamed life. Winged animals have been slaughtered by flying into turning turbine sharp edges. A large component of these issues has been settled or incredibly diminished through innovative improvement or by appropriately sitting wind plants.

3.4. HYDROPOWER

Hydropower is power gotten from the energy of falling water, running water, or sea energy (intensity of waves), which might be outfit for valuable purposes. It is an extremely regular resource which can be utilized to create power or to do helpful work.

Figure 3.3: A dam used to convert hydropower to electricity.

Source: https://upload.wikimedia.org/wikipedia/commons/7/73/Grand_Coulee_Dam.jpg

Hydroelectric power plants catch the energy of falling water to create power. A dam more often than not raises the water dimension of the river to make falling water, and it additionally makes it conceivable to power the progression of water (Figure 3.3).

The repository that is shaped is, essentially, put away energy. The power of falling water pushing against the turbine's cutting edges makes the turbine turn. The turbine changes over the active energy of falling water into mechanical energy.

A generator is associated with the turbine, so when the turbine turns, it makes the generator turn moreover. It changes over the mechanical energy from the turbine into electric energy. The amount of power a hydropower plant produces relies upon two variables: how far the water flows and the amount of water. Transmission lines lead power from the hydropower plant to homes and industries.

3.4.1. Conceivable Impact of Climate Transform on Resource Potential

The resource potential for hydropower is at present, dependent on verifiable information for the present climatic environments. With an evolving atmosphere, this resource potential could transform due to:

- Changes in river flow (overflow) identified with changes in neighborhood atmosphere, especially in precipitation and temperature in the catchment territory. This may prompt changes in overflow volume, changeability of stream and regularity of the stream (e.g., by changing from spring/summer high stream to more winter stream), legitimately influencing the resource potential for hydropower age.
- Changes in extreme events (floods and dry seasons) may increase the expense and hazard for the hydropower ventures.
- Changes in residue stacks because of changing hydrology and extreme occasions. Greater dregs could build turbine scraped areas and abatement effectiveness. Expanded silt burden could also top off supplies quicker and decline the live capacity, diminishing the level of guideline and diminishing stockpiling administrations.

3.4.2. Investment Cost of Hydropower Projects and Factors That Affect It

Fundamentally, there are two noteworthy cost groups for hydropower ventures:

- The construction costs, which ordinarily are the significant expenses of the hydropower venture; and
- The cost identified with electromechanical gear for energy change. Moreover, Investment costs incorporate the expenses of arranging, ecological effect investigation, authorizing, fish, and natural life relief, entertainment alleviation, authentic, and archeological moderation and water quality checking and alleviation.

The natural development costs pursue the value pattern of the nation where the undertaking will be created. On account of nations with economies on the move, the natural development costs are normally lower than in created nations because of the utilization of nearby work and neighborhood development materials. Natural development costs are dependably site explicit, for the most part, because of the inborn qualities of the geology, land environments and the development plan of the undertaking. This could prompt distinctive Investment cost and levelized expenses of power (LCOE) notwithstanding for ventures of a similar limit.

The expenses of electromechanical gear – as opposed to natural development cost – pursue world market costs for these segments. In hydropower ventures where the introduced limit is under 5 MW, the electromechanical hardware costs will in general command.

As the limit expands, the expenses are progressively impacted by the expense of natural buildings. The segments of the development venture that sway the natural development costs most are dams, admissions, water powered weight conductors (passages and penstocks) and power stations; in this manner, these components must be improved cautiously amid the building configuration arrange.

3.4.3. Operation and Maintenance (O&M) Costs

When assembled and put in operation, hydropower plants more often than not require next to no upkeep and activity expenses can be kept low, since hydropower plants don't have repeating fuel costs. O&M costs are generally given as a level of Investment cost per kW.

3.4.4. Small Hydro

Hydropower on a small scale is a standout amongst the most financially cost-effective energy advances to be considered for country charge in less created nations. It is also the fundamental prospect for future hydro advancements in Europe, where the extensive scale openings have either been abused as of now or would now be considered earth unsatisfactory.

Little hydro innovation is incredibly powerful (systems can keep going for a long time or more with small support) and is also a standout amongst the most earth kind energy advancements accessible.

3.4.5. The Challenges of Hydropower

- **Costly to Install and Maintain:** One can't disregard the way that the expense of development and upkeep of dams, power plants, and bolster foundation to deliver the power required for regular use can keep running into billions of dollars. This legitimately infers the mind-boggling expenses of form and upkeep can undermine the general generation cost of power.
- **Social Unrest:** Building dams on rivers and other water bodies will confine the availability of water to individuals who make their vocation through agribusiness and angling, in this way antagonistically influencing their salary.

More often than not, individuals got in such circumstances are left with no decision yet to migrate to different spots to make a decent living. This movement may realize social agitation among them because of the lost agricultural land.

- **Land Use:** The span of the store made for hydroelectric operations relies upon the amount of the hydroelectric generators and the geography of the land. In any case, flooding area to make a hydroelectric supply has antagonistic ecological effects as it crushes forests, natural life living spaces, and horticultural and grand grounds. In numerous examples, for example, the Three Gorges Dam in China, whole networks have must be migrated to manufacture supplies.
- **Impact on Aquatic Life:** Hydroelectric power plants can have many negative consequences for amphibian biodiversity. For instance, a few types of fish and different organisms can be harmed as well as murdered by turbine sharp edges.

- **Nature of Water:** Water in supplies is dormant rather than that in streams. Because of this, supplies will, in general, have larger amounts of residue and supplements than ordinary, which can offer ascent to the issue of over the top development of green growth and weeds to the degree that the weeds can swarm the river and leave practically no space for other sea-going widely varied vegetation.

Another challenge here is that water dissipates at a lot quicker rate in supplies than it does in rivers.

- **Greenhouse Gas Emissions:** Greenhouse gases (GHG) are discharged at the season of form and dismantling hydroelectric offices relying on the amount of the store. Notwithstanding, as per ongoing discoveries, these outflows don't stop there.

Critical amounts of outflows that add to a global warming are delivered amid the plant's operational cycle too (in spite of the fact that these dimensions are much lower contrasted with those created by coal-terminated plants).

- **Geological Limitations:** It is beyond the realm of imagination to expect to assemble a hydroelectric power plant at an arbitrary area as the plant will require explicit environments to be utilitarian. A dam should be developed over a water body which ought to have enough water to have the capacity to effectively create power. Subsequently, the development of such offices is principally subject to territories with adequate supply of water

3.5. GEOTHERMAL POWER

Geothermal power will be power separated from heat put away in the Earth. This geothermal energy starts from the first arrangement of the planet, from radioactive rot of minerals, and from solar energy retained at the surface.

A piece of direct geothermal heating limit is introduced for locale heating, spas, modern procedures, desalination, and agricultural applications. Geothermal power is financially cost-effective, dependable, practical, and ecologically well disposed.

The Earth's geothermal resources are hypothetically more than sufficient to supply mankind's energy needs, yet just an extremely small division might be productively abused. Boring and investigation for profound resources is over the top expensive (Figure 3.4).

Figure 3.4: Newberry geothermal project.

Source: https://farm4.staticflickr.com/3884/14513382913_d3cdf5535f_b.jpg

Geothermal electric plants were generally manufactured solely on the edges of structural plates where high-temperature geothermal resources are accessible close to the surface. The enhancements in penetrating and extraction innovation empower building geothermal systems over a lot more noteworthy geological range.

The heat productivity of geothermal electric plants is low, around 10–23%, in light of the fact that geothermal liquids don't achieve the high temperatures of steam. Fumes heat is squandered, except if it very well may be utilized directly and locally, for instance in nurseries, timber plants, fisheries, and area heating.

Direct heating is definitely more proficient than power age as it can utilize heat resources with lower temperatures. Where normal hot springs are accessible, the warmed water can be funneled directly into radiators.

There is a sure ecological hazard associated with the utilization of geothermal heat, in particular, high temp water from geothermal sources holds broke down gases and now and then small amounts of dangerous synthetic substances like mercury, arsenic.

Geothermal power requires no fuel (with the exception of siphons) and is subsequently safe to fuel cost variances; however, capital expenses are critical. Penetrating records for over a large component of the expenses and investigation of profound resources involves critical dangers.

3.5.1. Earth's Geothermal Energy Basics

The temperature at Earth's center estimates near 7200 degrees Fahrenheit. As one may envision, such a high temperature can deliver unprecedented amounts of economical energy and untold gigawatts of power. Actually, geothermal energy is viewed as an inexhaustible source of energy which can create energy for whatever length of time that the planet exists.

As per the Geothermal Energy Association (GEA), the geothermal power industry came to around 3,442 MW toward the finish of 2013. Add to this around 1,000 MW of arranged limit options a work in progress and around 3,100 MW of geothermal energy resources being worked on. The business' US options in 2013 included around 85 MW of new limit at new and repaired power plants in Utah, Nevada, California, and New Mexico.

The GEA's 2014 Annual U.S. and Global Geothermal Power Production Report expresses, "The universal geothermal power showcase is blasting, developing at a continued rate of 4% to 5%. Right around 700 geothermal operations are being worked on in 76 nations.

Numerous nations foreseeing the dangers brought about by environmental transform understand the estimations of geothermal power as a baseload and some of the time adaptable source of sustainable power source."

Upper assessments demonstrate a wonderful worldwide geothermal energy capability of 2 terawatts (TW); however, considerably more than humans are on track to tap. At that point, for what reason aren't humans delivering such energy? How about humans investigate a component of the points of interest and disservices with regards to creating geothermal energy for home, business, or city.

3.5.2. Geothermal Energy Advantages

- **Environmentally Cordial:** There are a couple contaminating perspectives to tackling geothermal energy, and the carbon impression of a geothermal power plant is viewed as negligible. A normal geothermal power plant discharges what might be compared to 122 kg CO_2 for each megawatt-hour (MWh) of power it creates – one-eighth of the CO_2 emanations related with natural place coal power plants.

- **An Inexhaustible Resource:** Geothermal stores are normally recharged. As indicated by certain researchers, the energy in the geothermal supplies will last billions of years. While non-renewable energy sources have an expiry date, inexhaustible sources like geothermal energy won't terminate at any point in the near future.
- **Potential Limit:** Estimates for the capability of geothermal power plants fluctuate between 0.035 to 2 TW.
- **A Steady Resource:** The power yield of a geothermal plant can be precisely anticipated. Not expose to a similar low-energy changes also with solar or wind.
- **Great for Heating/Cooling:** There is huge development in the quantity of mortgage holders using geothermal heating/cooling throughout the most recent few years.
- **No Fuel Required:** After establishment, no mining or transportation action is essential.
- **Small Land Impression:** Of any significant power source.
- **Stable Resource:** Can give base burden or pinnacle control.
- **Economic Elements:** Cost-focused in certain zones.
- **Accessibility:** Some dimension of geothermal energy accessible generally puts.
- **Renewable:** Geothermal energy is separated from earth's center and will be accessible insofar as earth exists. It is in this way inexhaustible and can be utilized for generally another 4–5 billion years.
- **Abundant Supply:** With geothermal energy, there are no deficiencies or different sorts of issues which now and again happen with different kinds of intensity.
- **Significant Savings for Home Owners:** There has been a colossal increment in the quantity of property holders who need to use geothermal energy for heating and cooling purposes. The outcome is that less energy is utilized for heating homes and workplaces, which results in critical investment funds for mortgage holders. After the underlying cost, a 30–60% investment funds on heating and 25-half reserve funds on cooling can take care of that expense inside couple of years.

3.5.3. Geothermal Energy Disadvantages

- **Potential Outflows:** Greenhouse gas underneath Earth's surface can possibly move to the surface and into the environment. Such emanations will, in general, be higher close geothermal power plants, which are related with sulfur dioxide and silica discharges. Additionally, and the stores can contain hints of poisonous substantial metals, including mercury, arsenic, and boron.
- **Surface Instability:** Construction of geothermal power plants can influence the solidness of land. In January 1997, the development of a geothermal influence plant in Switzerland set off a quake with an extent of 3.4 on the Richter scale.
- **High Expense for Power:** Total expenses generally end up somewhere close to \$2–7 million for a 1 MW geothermal power plant.
- **High Straightforward Expenses for Heating and Cooling Systems:** While there is an anticipated ROI, it won't occur rapidly. For a normal estimated home, establishment of geothermal heat siphons costs between \$10,000 – \$20,000 which can satisfy itself in another 5–10 years down the line.
- **Location Specific:** Good geothermal stores are difficult to find. Iceland and Philippines meet almost 33% of their power request with geothermal energy. Prime destinations are regularly a long way from populace focuses.
- **Distribution Costs:** If geothermal energy is transported long separations, cost can wind up restrictive.
- **Sustainability Questions:** Some investigations demonstrate that stores can be exhausted if the liquid is expelled quicker than replaced. This isn't an issue for private geothermal heating and cooling, where geothermal energy is being utilized uniquely in contrast to in geothermal power plants.
- **Cost of Powering the Pump:** Geothermal heat siphons need a power source.
- **May Run Out of Steam:** One must be unbelievably cautious when one is endeavoring to check everything that is identified with geothermal energy. Mind must be taken to watch the heat and not to manhandle it, supposing that the heat isn't dealt with appropriately, it can cause an emergency or different issue where

the energy isn't legitimately dispersed or utilized.

Regardless of whether geothermal energy will be utilized for heating a house or creating power at a power plant, there is a lot to be taken a gander at and considered. In spite of the fact that the impression for this type of energy is extensively smaller than one than that of petroleum derivatives, despite everything it has conceivable negative effects.

At that point, there is the issue of expense. Geothermal power plants are costly to make and not generally that simple to discover. Also, the underlying expense for buildings or houses utilizing a geothermal heating and cooling system is costly and without a quick compensation.

3.6. BIOMASS

Biomass is natural material gotten from living or as of late organisms. Humans as people make an enormous number of side-effects, including human waste, general waste, animal waste, and significantly more.

Biomass transformation innovations presently exploit these and by consuming side-effects, they may discharge the energy legitimately, as heat or power, or may transform over it to another form, for example, fluid bio-fuel or flammable biogas, rather than further contaminating the air.

Many waste treatment offices and landfill locales are moving to biomass energy to make power and to make their towns and urban communities cleaner (Figure 3.5).

Figure 3.5: Biofuel plant in Sweden.

Source: https://upload.wikimedia.org/wikipedia/commons/c/cd/FIL2938.JPG

3.6.1. Classes of Biomass Materials

There are five fundamental classifications of material:

- Virgin wood, from ranger service or from wood handling;
- Energy crops: high return crops developed explicitly for energy applications;
- Agricultural deposits: buildups from agribusiness reaping or handling;
- Food squander, from food and drink production, arrangement, handling, postconsumer squander; and
- Industrial waste and co-goods from assembling and modern procedures.

As there is wide diverse variety in the attributes and properties of these distinctive classes of material and their different sub-collections, there is additionally a wide scope of transformation advances, which incorporate both heat (ignition, gasification, pyrolysis) and substance transform advances (anaerobic processing, aging, treating the soil).

Biomass power plant estimate is frequently determined by biomass accessibility in closeness as transport expenses of the (massive) fuel play a key factor in the plant's financial aspects.

The current challenges of biomass inventory network identified with various feedstock can be extensively arranged into operational, monetary, social, and approach and administrative challenges.

3.6.2. Operational Challenges

- **Feedstock Inaccessibility:** Inefficient resource the board and the administration non-intercession approach are the key factor frustrating the extension of the biomass business.
- **Local and Regular Accessibility of Biomass and Capacity Issue:** The occasional variety results in the fuel cost. As the energy thickness of biomass is low, procurement of land for collecting and capacity is troublesome.
- **Weight on Transport Area:** Because of biomass dampness, transporting wet biomass from the manor to the generation site

turns out to be enthusiastically ominous and expensive with expanding separation.

- **Wastefulness of Transformation Office, Center Innovation and Hardware Deficiency:** Technical obstructions were come about because of the absence of benchmarks on bioenergy systems and gear, particularly where the energy sources are so various. Suitable pretreatment required to avert biodegradation, and loss of heating quality expands the generation cost as well as in gear's venture.
- **Juvenile Industry Chain:** It is for all intents and purposes difficult to get long haul contracts for predictable feedstock supply in sensible cost. The low capacity to pick up benefits is additionally a reason that numerous upstream firms need main thrusts in the innovation change.

3.6.3. Financial Challenges

- **Feedstock Procurement Cost:** The biomass resources are dissipated, and so as to diminish the expense of transportation, biomass ventures are anxious to possess land near the source, prompting centralization of biomass ventures.
- **Restricting Financing Channels and High Investment and Capital Expense:** Because of decentralized capital, poor productivity, visit vacillations of worldwide raw petroleum costs and high market chance, rare speculators showed a drive part in the biomass power age industry. The biomass power age is exposed to imperatives of unreasonable venture and high working expenses. Biomass pretreatment innovations have additional costs, which dissipated ranchers and small-scale fuel organizations will be unable to manage.

3.6.4. Social Challenges

- **Clashing Choice:** Decision making on choice of provider, area, courses, and advances is pivotal and needs legitimate correspondence. By fortifying collector and actualizing the duties, the partners ought to be made completely mindful of the financial, ecological, and social abundance of resource usage.
- **Land Use Issues:** It prompts the loss of biological systems safeguarding and the homes of indigenous individuals.

- **Effect on Nature:** The biomass estate exhausts supplements from soil, advance tasteful debasement, and increment the loss of biodiversity. Other social effects will result from establishment of energy cultivates inside rustic territories like expanded need of administrations, expanded traffic, and so forth. The potential negative social effects seem sufficiently able to disregard the advantage of new and perpetual work age.

3.6.5. Arrangement and Regulatory Challenges

- **Arrangements:** At present, the administration is sponsoring the local fuel value, which thus makes the power creating cost from traditional sources lower than the power generation cost from inexhaustible energies.
- **System:** There are no particular principles to manage crafted by use of biomass resource, and there are no particular punishments for not utilizing conduct that ought to be thoroughly utilized.
- **Guideline:** There is no unique component to deal with the improvement of biomass resources industry and no specific office to deal with the usage of important national gauges and arrangements.

3.7. FRESH WATER

Water can be viewed as an inexhaustible material in states of painstakingly controlled utilization, treatment, and discharge. If not, it would turn into a non-inexhaustible resource at that area. For instance, groundwater could be expelled from an aquifer at a rate more noteworthy than the reasonable energize. Expulsion of water from the pore spaces may cause changeless compaction that can't be recharged (Figure 3.6).

Figure 3.6: Freshwater can be used to produce good amounts of energy.

Source: https://farm1.staticflickr.com/40/111011129_f63ef2f2ed_b.jpg

3.8. FORESTS

Forests are normally inexhaustible if expulsion of the trees is controlled. Humans are seeing the procedure of deforestation (decimating or expelling a woodland biological system) in numerous territories currently. Trees are cut for fuel, for the benefit, for obtaining area to plant crops or develop animals.

Once in a while, deforestation is normal (substantial flames). Species may end up wiped out (the loss of their living space), disintegration, and flooding can happen just as desertification and diminished land efficiency. The general atmosphere of a region can also change. Numerous nations have created programs gone for diminishing deforestation (Figure 3.7).

Figure 3.7: Forests supplying wood as a resource.

Source: https://www.blm.gov/sites/blm.gov/files/naturalresources_forest-stewardshipcontracting1_2.jpg

3.8.1. Planted Forests a Favored Source of Sustainable Power Source

Wood fills, including kindling and charcoal, are the most imperative energy source and the most essential forests item for some creating nations. All inclusive, about component of the yearly wood expulsions (about 1.7 billion m^3) are utilized for fuel.

Quickly developing planted woodlands can be a proficient method to deliver wood fuel. Right now, they supply an expected 15 to 20% of the world's wood energizes, regardless of whether as results or buildups from modern forests estates or as entire trees from committed bioenergy farms (Populus, Salix, Eucalyptus, Acacia), which by and large don't contend with food generation as they develop ashore that is unsatisfactory for serious agribusiness.

However, the evaluation of the effects of bioenergy farms is mind-boggling, and there are various potential dangers for food security, land use transforms, and the transformation of significant eco-systems.

The utilization of sustainable power source resources, including the utilization of wood and woody biomass is progressively advanced by industrialized nations and request is probably going to increment further because of high costs of petroleum product options.

The congress focused anyway that satisfying this expanding need is tricky in light of the fact that there is deficient region of land reasonable for planted forests. Further, the all-out supply of wood fuel isn't accessible for bioenergy generation, as in numerous nations this resource is imperative for heating and cooking. In this unique situation, the congress noticed that the challenge between the material and energy utilization of wood is heightening and should be enough overseen.

3.8.2. The Management Intercessions to Adapt to Environmental Change

The essentialness of afforestation and reforestation in adjustment to, and moderation of environmental transform is broadly recognized in the instruments of the UN System Convention on Climate Transform (Clean Development Mechanism, REDD).

In this setting, the congress featured that the arrangement of woodland products and biological system administrations from planted forests will be adjusted by environmental transform marvels presenting new challenges to forests organizers and chiefs.

Explicit mediations in planted forests the management to adapt to this circumstance includes:

- Recognize, overview, and amount existing woodland resources and their most ideal uses so as to broaden the board, and to diminish the danger of harming environmental transform impacts;
- Screen normal environments to acquire early signs that changes are happening, and to encourage compelling reactions to such changes;
- Plant tree species that are versatile to anticipate environmental transform impacts on collected, consumed or corrupted woodland land and relinquished Agriculture area;
- The other way around, counteract deforestation of timberland land;
- Build up short-pivot woody biomass estates on reasonable destinations to substitute bioenergy for non-renewable energy sources;
- Differentiate buildings and species in planted woods to expand their versatile ability to climatic risks;
- Build up mixed species estates, don't substitute natural forests for

planted forests, lessen the size of group operations and improve the network among characteristic and planted woodland zones;

- Keep up wellbeing and imperativeness of planted forests by actualizing flame, bug, and sickness counteractive action amounts and by applying fitting silvicultural medicines;
- Diminish outflows from deforestation and woodland corruption through incorporated flame and vermin the board, the selection of low effect collecting strategies and stretch out pivot periods to lessen harms to soil ripeness and standing trees;
- Oversee and protect the natural hydrological cycle on a scene level to keep up or alter the accessibility of water for tree and plant development.

The congress concurred that there was a need to help nations in arranging and actualizing these intercessions and to help them in understanding the job that planted forests can play in adjustment to and moderation of environmental change.

3.8.3. Effects of Planted Forests on Organic Diverse Variety

The Congress perceived that the protection of natural decent variety and hereditary resources is one of the key issues in forests the board. The positive or negative effect of planted woodlands on organic decent variety depends to a substantial degree on the environment and past utilization of the land being planted and on the power of the board connected at a scene level.

The form of planted forests needs to grasp standards of multifaceted nature, versatility, and self-guideline for biodiversity preservation.

Huge scale single species estates with solid ecological effects ought to be replaced by increasingly biological and incorporated methodologies at stand and scene levels, including mixed species manors, more noteworthy utilization of silvicultural systems dependent on near nature standards, preservation the management of regular woodland remainders and the planting of local species just as presented species.

The congress noted further that the capability of local species is regularly ignored in planted forests since a component of the previous may accomplish development rates practically identical to those of presented species given appropriate silvicultural the board and cautious hereditary determination.

Presented species ought to be chosen just in connection to explicit administration goals, economic situations and natural site environments and the hazard that these species may wind up obtrusive and effectively affect the neighborhood biodiversity ought to be assessed.

3.9. AGRICULTURE GOODS

Supportable horticulture represents the utilization of systems which take into account insignificant and controlled ecological harm. Goods from this sort of agribusiness are inexhaustible and manageable when preparing, and coordination identified with these goods also have economical attributes.

Biomass energy sources on the farm are given as underneath.

3.9.1. Biomass Residues

Agriculture exercises produce a lot of biomass deposits. While most yield buildups are left in the field to lessen disintegration and reuse supplements once again into the dirt, some could be utilized to create energy without hurting the dirt.

Different squanders, for example, whey from cheddar generation and compost from domesticated animals' operations can also be beneficially used to create energy while diminishing transfer expenses and contamination.

3.9.2. Energy Crops

Harvests developed for energy could be created in huge amounts, similarly as food yields may be. While corn is right now the most generally utilized energy crop, local trees and grasses are probably going to turn into the most mainstream later on.

These enduring harvests require less upkeep and fewer contributions than do yearly column crops, so they are less expensive and increasingly maintainable to create.

3.9.3. Grasses

Switchgrass has all the earmarks of being the most encouraging herbaceous energy crop. It creates significant returns and can be reaped every year for quite a long while before replanting. Other local varieties that develop rapidly, for example, enormous bluestem, reed canary grass, and wheatgrass, could also be beneficial.

3.9.4. Trees

Some quickly developing trees make amazing energy crops, since they develop back more than once in the wake of being sliced off near the ground. These short-turn woody yields can develop to 40 feet in under eight years and can be collected for 10 to 20 years before replanting. In cool, wet locales, the best decisions are poplar and willow. In hotter territories, sycamore, sweetgum, and cottonwood are ideal.

3.9.5. Oil Plants

Oil from plants, for example, soybeans, and sunflowers can be utilized to make fuel. Like corn, however, these plants require more serious administration than other energy crops.

3.9.6. Securing the Land

With keen practice and the management, perpetual energy yields can improve the dirt nature of land that has been abused for yearly column crops. The profound foundations of energy crops upgrade the form of the dirt and increment its natural substance.

Since working happens rarely, the dirt experiences minimal physical harm hardware. One examination assesses that changing over a corn ranch of normal size to switchgrass could spare 66 truckloads of soil from disintegration every year.

Enduring energy crops need impressively less compost, pesticide, herbicide, and fungicide than yearly column crops. Diminished concoction uses shields ground and surface water from toxic substances and over the top amphibian plant development.

Moreover, profound established energy harvests can fill in as channels to shield conduits from synthetic overflow from different fields and avert sedimentation brought about by disintegration.

At long last, enduring energy harvests can make more different habitats than yearly line crops, drawing in a more extensive variety of species, for example, winged animals, pollinators, and other helpful creepy crawlies, and supporting bigger populaces. Moreover, the long collect window for energy crops empowers ranchers to abstain from settling or reproducing seasons.

3.9.7. Changing Over Biomass to Energy

Most biomass is changed over to energy a similar way it generally has been—by consuming it. The heat can be utilized legitimately for heating buildings, crop drying, dairy activities, and modern procedures. It can also be utilized to deliver steam and produce power.

For instance, numerous electric generators and organizations consume biomass independent from anyone else or with different fills in traditional power plants.

Biomass can also be changed over into fluids or gases to create power or transportation fills. Ethanol is normally delivered through aging and refining, in a procedure much like that used to make brew. Soybean and canola oils can be synthetically changed over into a fluid fuel called biodiesel. These powers can be utilized in ordinary motors with close to nothing, assuming any, and change.

Biomass can be changed over into a gas by heating it under strain and without oxygen in a "gasifier." Manure also can be changed over utilizing a digester. The gas would then be able to be signed to deliver heat, steam, or power.

Various biogas applications are still being developed; however, show incredible potential. One promising innovation is immediate ignition in a propelled gas turbine to run a generator and produce power. This procedure is twice as proficient as basically consuming crude biomass to deliver power from steam.

Analysts are also growing little, fast generators to keep running on biogas. These "microturbines" have close to three moving parts and create as meager as 30 kilowatts, which could power a medium-sized homestead. A few organizations are also considering changing over gasified biomass into ethanol as a more affordable option in contrast to maturation.

On the other hand, biogas can be prepared into hydrogen or methanol, which would then be able to be artificially changed over to power in an exceedingly proficient energy unit. Energy components can be sufficiently substantial to power a whole ranch or small enough to power a vehicle or tractor.

An inventive examination in Missouri gives one case of the conceivable outcomes. Corn is utilized to create ethanol, and the loss from the procedure is nourished to cows for a dairy generation. Dairy animals' excrement treats the corn and is additionally gone through a digester to create biogas.

An energy component productively changes over the biogas into power to run the operation. The final results are ethanol, power, and milk. All the waste goods are utilized inside the undertaking to bring down expenses.

REFERENCES

1. Algburi, S., (2017). *Hydroelectric Power*. [eBook] Available at: https://www.researchgate.net/publication/315493188_Hydroelectric_Power (Accessed on 21 May 2019).
2. Energy.gov., (n.d.). *Advantages and Challenges of Wind Energy*. [online] Available at: https://www.energy.gov/eere/wind/advantages-and-challenges-wind-energy (Accessed on 21 May 2019).
3. Eubia.org., (n.d.). *Challenges Related to Biomass – European Biomass Industry Association*. [online] Available at: http://www.eubia.org/cms/wiki-biomass/biomass-resources/challenges-related-to-biomass/ (Accessed on 21 May 2019).
4. Mara, E., (2014). *Top Challenges in Using Hydroelectric Power*. [online] Engerati – The smart energy network. Available at: https://www.engerati.com/blogs/top-challenges-using-hydroelectric-power (Accessed on 21 May 2019).
5. Meyers, G., (2016). *Geothermal Energy Advantages and Disadvantages*. [online] Planet save. Available at: https://planetsave.com/2016/02/11/geothermal-energy-advantages-and-disadvantages/ (Accessed on 21 May 2019).
6. Need.org., (2018). *Solar*. [online] Available at: https://www.need.org/files/curriculum/infobook/solari.pdf (Accessed on 21 May 2019).
7. *Planted Forests are a Vital Resource for Future Green Economies*, (2013). [eBook] FAO. Available at: http://www.fao.org/forestry/37902-083cc16479b4b28d8d4873338b79bef41.pdf (Accessed on 21 May 2019).
8. *Renewable Resources*, (n.d.). [eBook] Available at: https://sc-s.si/joomla/images/Renewable%20resources.pdf (Accessed on 21 May 2019).
9. Ugc.ac.in., (2004). *Natural Resources*. [online] Available at: https://www.ugc.ac.in/oldpdf/modelcurriculum/Chapter2.pdf (Accessed on 21 May 2019).
10. Union of concerned scientists, (2003). *Growing Energy on the Farm: Biomass Energy and Agriculture*. [online] Available at: https://www.ucsusa.org/clean-energy/increase-renewable-energy/biomass-energy-agriculture (Accessed on 21 May 2019).

4

Renewable Energy Sources

CONTENTS

Renewable energies are emerging as a preferred alternative for generating power. They are produced by harnessing the renewable sources that are abundantly available on Earth. They are distributed across the world making it accessible to all countries. The emerging economies, in particular, will benefit immensely when these sources are adopted. This chapter explains the concept of renewable energy. It then elucidates the sources it is obtained from. There are multiple sources, sunlight, and wind being omnipresent. This makes multiple sources accessible to all countries. This chapter then discusses the significance of renewable sources of energy. This ranges from mitigation of climate change to inexhaustible supply to reducing the vulnerability of energy systems amongst others. The chapter also discusses the market development of various sources and its applicability. Sustainable development is a recurring theme in all major initiatives at a global level. The role of renewable sources of energy in ensuring sustainable development, economic well being, and energy security is also highlighted in this chapter. Finally, the future scenario for renewable sources of energy is covered in this chapter.

4.1. INTRODUCTION

The planet Earth contains many resources, both renewable and non-renewable. Renewable resources refer to those that can be replenished as it gets consumed. Generally, they can be replaced at the same or higher rate at which they are drawn from the Earth. Some of the resources are considered as infinite that which has an endless supply.

Some examples of infinite renewable resources are wind energy, solar energy, and geothermal pressure. While there are others which though renewable, take some time and involves efforts. If they are replenishing at a rapid rate, it will be difficult to balance the demand and supply. Some of these resources are wood, seafood, vegetation, etc. Precious metals can also be considered as renewable resources. Although they cannot be replenishing, they can be recycled. Precious metals do not lose their properties when they are extracted and used, making recycling possible.

The population on Earth is growing exponentially. As a result, there is an increasing demand for all resources, which makes renewable resources even more crucial. There are many resources like the fossil fuels that have a finite supply. Generating energy from renewable resources puts less strain on the limited supply.

However, at present energy from renewable resources are not used on a large scale. More advanced technologies are needed to use these resources in an efficient manner. Progress has been made in this field. The energy generated from non-hydroelectric renewable energy and bio-fuels has more than doubled from 2000 to 2017. This can be largely attributed to initiatives taken by various governments across the world like requirements imposed and other tax benefits.

Although all energy resources are important, renewable resources can be replaced after use, which makes it more potent to address sustainability issues. Whereas, non-renewable resources can be used only once. Hence the governments are directing their efforts towards conservation of nonrenewable resources while trying to increase the supply of renewable energy sources. One of the leading causes of environmental issues and scarcity of energy resources is overpopulation.

The natural resources are the biggest assets for Earth. They are also known as natural capital. In recent years, more awareness about sustainability has made the alternate energy sources even more valuable. They have an advantage over non-renewable resources like oil, natural gas, coal which are exhausted once consumed. The renewable sources of energy are expected to act as a suitable substitute for non-renewable energy resources.

The challenge at present is that the prices for renewable energy resources are higher. However, with the advancement in technology, these challenges can be overcome in the near future. Moreover, the decrease of supply on nonrenewable sources of energy compared to its demand is expected to increase their price. The nonrenewable resources will then emerge as a cost-effective alternative.

The renewable resources of energy include wind, water, solar, and bio-fuel. For example, windmills are used to generate energy by harnessing the wind power. Biodiesel, a type of biofuel, is a substitute for oil. Another example of bio-fuel is the green diesel, which is made from algae and other similar plants.

The renewable energy resources have a significant global impact. Renewable resources have become a focal point of the environmental movement, both politically and economically. In the long run, the renewable energy resources are indispensable. They can effectively substitute the fossil fuels which cannot be replenished once used. As mentioned earlier, these resources are not yet cost-effective and require further research for using them efficiently.

Adopting sustainable energy is often referred to as "going green" due to the positive impact on the environment. It not only addresses the issue of resource scarcity, it also mitigates global warming. The renewable resources of energy do not cause damage to the environment, unlike the burning of fossil fuels, which emits greenhouse gases (GHG).

The Kyoto Protocol, which was signed in 1997, was the first major international accord with the purpose of reducing carbon dioxide emissions and curbing global warming. In 2015, the global powers held a meeting in Paris to increase the adoption of renewable resources of energy and curbing the emissions of GHG.

Countries all over the globe have taken initiatives to encourage adoption of renewable source for generating energy. For instance, a surcharge has been levied on fossil fuels, which make the renewable resources more competitive. This will make people chose renewable energy sources instead of fossil fuels. Green funds investment vehicles such as mutual funds, support eco-friendly, and sustainable companies by investing in them and helping to promote environmental awareness. The most important step is increasing awareness amongst the public and making them aware of the impact of using renewable sources of energy.

The initiatives taken by the government have seen considerable success. As per data provided by the U.S Energy Information Administration (EIA), almost 11 quadrillion British thermal unit can be attributed to the use of renewable energy sources. This represents 11% of the entire energy consumption in U.S. Around 57% of U.S. renewable energy in 2017 was consumed by the electric power sector, and approximately 17% of U.S. electricity generation was from renewable energy sources.

The State and Federal governments have taken many steps to replace the use of fossil fuels. These include imposition of requirements and incentives for adopting bio-fuels. It is anticipating that the current growth trend will continue in the coming years, and the use of renewable energy resources will reach an optimal level by 2050 (Figure 4.1).

Figure 4.1: Representation of renewable energy park.

Source: https://upload.wikimedia.org/wikipedia/commons/3/32/Renewable_energy_park.jpg

4.2. TYPES OF RENEWABLE ENERGY SOURCES

Renewable energy sources are literally found in our surroundings in the sunlight, wind, oceans, and inside Earth's crust. They are components of the physical structure of Earth and are being replenished continually on its own. Hence, they do not carry the risk of being exhausted.

These sources are referred to as 'alternative energy' since they are substitutes to conventional fossil fuels like coal and oil. It is important to understand that renewable energy sources can guarantee a continuous supply. However, it does not make it completely environmentally safe. For instance, water dams built to harness water energy can have a negative impact on the water animals. Wind turbines can generate clean energy using solar PV. However, the generation can have some negative impact on the environment.

However, the environmental impact caused by clean technologies is much lower than the usage of fossil fuels. This makes renewable energy sources indispensable. They can help in combating the pollution problem. This can help manage climate change in the long run. Besides being eco-friendly, they are also good for domestic and business purposes.

Much advancement has been made in the field of clean technologies. In many parts of the world, solar, and wind energy is cheaper than the conventional sources. The price keeps dropping further. At present, there are seven renewable energy sources which can be harnessed to generate power. Except for Hydrogen and Geothermal energy source, the sun is a source of energy for the rest of the clean energy sources.

4.2.1. Solar Energy

Sunlight is the most popular renewable energy resource. Sun is the source of energy for the entire planets, and all ecosystems depend on it for its existence. Solar energy generation has no environmental impact as it does not release pollutants; neither use fossil fuels. Besides, the cost of this technology is also reducing increasing its worldwide acceptance.

The sun's energy is captured directly and is used to generate power. Various technologies are used to harness the sun's energy and transform the light into heat. It can then be used for lighting, heating, and generating electricity. Cooling systems for commercial purposes can also work on solar energy.

Photovoltaic (PV) systems have solar cells embedded in them, which can capture the sunlight and transform it into energy. The solar hot water system can be used to heat buildings by circulating water through flat-plate solar collectors. It makes use of a technology similar to conventional steam generator. The mirrored dishes in the solar panel can generate electricity by concentrating the sun's energy.

Solar energy can also be used to meet large scale requirements like in case of industrial and commercial buildings. It can provide ventilation, cooling, and heating to these buildings by leveraging sun's energy. Design of the solar PVs is an important aspect which can leverage the sun as a source of light and use it for heating as well as cooling.

Solar power can be harnessed to suit the domestic and commercial purposes in many ways. Solar panels are designed to meet specific purposes of domestic or commercial use. A building can have an inbuilt solar system at the time of construction, or it can be fitted even old buildings that have already been constructed. Cooling and ventilation systems installed can use sunlight as a source of energy. Systems can be designed from scratch structures that take advantage of the sun's natural attributes for passive heating and lighting (Figure 4.2).

Figure 4.2: Solar energy panels are clean technologies.

Source: https://farm9.staticflickr.com/8065/28905292973_9cfa708c59_b.jpg

4.2.2. Wind Energy

Wind is another form of energy which available abundantly. The wind is caused by movement of air as a result of uneven heating of the Earth's surface and the rotation of the Earth. Wind energy is harnessed using huge turbines which generate power when they spin.

Wind energy has had limited application for individual households. However, it is emerging as a popular technique for utility scale application. On a small scale, windmills are used for irrigation purpose. Massive wind spread over several square miles is found across the world. Wind energy is also a clean source like the solar energy. It is emerging as an increasingly important renewable energy source across the world.

Commercial grade wind-powered generating systems are available to meet the renewable energy needs of many organizations. Single-wind turbines can be used as an add-on for an existing conventional source of energy. When wind energy is available, the system offsets the need for conventional energy source. Utility-scale wind farms can be used to generate massive amounts of power. It can be acquired from the wholesale power market by a bidding process or by signing a contract.

4.2.3 Geothermal Energy

Geothermal energy is generated from the heat of the Earth. The heat can be sourced from the surface or miles deep into the crust from heated rocks and hot water reservoirs. Geothermal power plants generate power by leveraging the heat sources. It can also be used on a small scale using a geothermal heat pump system. It can harness the heat from ten feet underground. It can be used for both heating and cooling as per the requirement.

Geothermal energy can be part of a commercial utility energy solution on a large scale or can be part of a sustainable practice on a local level. Some direct uses of geothermal energy are heating office spaces and manufacturing facilities, supplying power to greenhouse plants, maintaining water temperature in pisciculture and support various other industrial processes like pasteurizing milk.

Geothermal energy comprises a small percentage of the energy mix. However, it has great potential and can be harnessed to supply a massive amount. Iceland has captured this potential well and almost 90% of the home heating needs, and 25% of the electricity needs is met by geothermal energy. There are some drawbacks of this source like it involves huge cost for constructing power plants; it is affected by the instability of Earth's surface like Earthquakes.

4.2.4. Hydroelectricity

Similar to water movement, water currents can also generate energy. This renewable energy is generated when water is passed through a turbine which spins it to generate electricity. Water current is naturally formed in waterfalls and dams when water drops from a significant height. This natural phenomenon is leveraged at the Niagara Falls and Hoover Dam on the Colorado River.

Hydropower is a clean source of energy as it does not emit any pollutants or GHG. However, compared to other renewable sources of energy, hydroelectricity has a greater environmental impact. They can cause a change in water levels; affect the water current change migration path for aquatic animals.

Hydropower has been around since historical times. Waterwheels were used to operate sawmills and gristmills in the earlier times. However, the technology has evolved and is captured in a different manner to generate electricity. This is known as "pumped-storage hydropower," where water is cycled between lower and upper reservoirs to control electricity generation between times of low and peak demand.

"Run of river Hydropower" is another method which uses a channel to funnel a part of the water flow and is not dependent on dam. It can be built according to various capacities from large scale projects to micropower systems (Figure 4.3).

Figure 4.3: Dams are a source of natural hydroelectricity.

Source: https://upload.wikimedia.org/wikipedia/commons/6/6c/Chief_Joseph_Dam.jpg

Direct use of hydroelectricity is possible only when a water source is geographically suitable to produce it naturally. When a dependable source is identified, which can supply uninterruptedly, micro-hydroelectric plant can be constructed. The power generated can be used to support ranch and farm operations and run small municipalities. Mid-sized hydroelectric power systems can be built to cater to the requirements of small rural towns. '

Although renewable energy sources are yet to become a significant proportion of the overall electricity mix, hydroelectricity is emerging as a leading renewable alternative. Besides large projects like the Hoover Dam, lower dams and underwater turbines can be constructed on small water bodies. This can be used to drive smaller projects.

4.2.5 Ocean Power Source

The ocean as a source of energy contributes to the generation of energy in two ways. The thermal energy from the sunlight and mechanical energy produced as a result of tides and waves.

The thermal energy of the ocean can be used to generate electricity using technology that relies on the heat generated from the water surface. Even the cold water deep inside the ocean, which does not receive sunlight, can be leveraged to produce a cooling effect. This method produces desalinated water as a by-product.

The other source is the mechanical energy generated by the oceans. It leverages the tidal energy that is caused because of gravitational effect of the moon and the rotation of the earth. Tides are predictable, which makes tidal energy one of the reliable sources. Tidal energy has been in use since the ancient times.

Usually, tidal energy has faced from relatively high cost as well as incomplete accessibility of sites through suitably high tidal ranges. The largest tidal energy power plant is in France and is called "Rance Tidal Power Station." The first marine energy center of the world was established in 2003 in Scotland and Orkney in 2003. This was setup to develop wave energy and tidal energy. Energy produced by waves can also be harnessed and used as a supplementary source of electricity. Several other technologies that are not that popular can harness ocean winds, ocean current, and saline properties as a means of power generation. More than 70% of the Earth's surface is covered by oceans; hence, it is emerging fast as a substitute of conventional energy. The future for energy harnessed from oceans look promising. It is influenced on geographies and regulations by the government. Seaside communities can employ the methods to tap natural ocean energy described above to supplement municipal power and energy needs.

4.2.6. Other Alternative Energy Sources

The above-mentioned sources are occurrence of natural processes. Two other types of renewable sources exist that have to be harnessed using mechanical means.

- **Bio-Energy:** It is a renewable energy which is produced from biomass. It is used to generate heat and electricity, as well as liquid fuels like biodiesel and ethanol, which are used in road vehicles.

Biomass is basically the organic remains of plants and animals. Bio-energy emits carbon dioxide like the fossil fuels. However, replacement plants are grown as biomass, which absorbs the CO2 that is emitted and mitigates any environmental impact.

Bio-energy is generated using various methods. This includes Combustion of biomass, capturing methane gas that is released when organic material decomposes and capturing them for use. Bio-energy is used mainly by vehicles. Transport business companies can convert their vehicles and make them compatible for using biofuels like biodiesel and ethanol. Manufacturing facilities can be equipped to burn biomass directly to produce steam captured by a turbine to generate electricity.

In some bio-energy can be used for supplying electricity to a facility as well as heating it up. One such example is the use of wood waste by paper mills to generate electricity as well as produce steam. Small modular systems generate electricity from the waste of livestock and use it for farming operations. Small towns can be electrified by bio-energy. It can tap the methane gas that is released by microorganisms during anaerobic digestion from the organic wastes which can be used as fuel.

- **Energy from Hydrogen:** Hydrogen is the simplest element and present everywhere in the universe. However, it is not present as a natural gas on Earth. It is present in organic compounds in hydrocarbon form such as gasoline, natural gas, propane. It is also an element in the water (H_2O). Certain bacteria and algae release hydrogen with the aid of sunlight under certain conditions.

Although hydrogen possess high amount of energy, it releases no pollutants when combusted. Since the year 1950, liquid hydrogen has been used in rockets and space shuttles for sending them to the space. Hydrogen fuel cells convert the potential chemical energy of hydrogen into electricity, with pure water and heat as the only byproducts.

However, widespread use for domestic and commercial purpose will only be possible will advancement in technology that can reduce the cost and improve its performance. At present, in the US, it is used to refine petrol, treat metals, food processing, and production of fertilizers. Besides, it is also used as a source of energy and is combined with oxygen to produce electricity.

Hydrogen-powered vehicles have been invented and are currently operating in United States. They will become more popular when the number of refueling station increases, and the price for the fuel comes down. There are other practical applications for hydrogen generated energy. It includes large fuel cells that can be used as back up for residential areas, electric motor vehicles that derives its power from hydrogen and marine vessels powered by the same technology.

4.3. IMPORTANCE OF RENEWABLE ENERGY

Infinite supply of renewable is the most significant attribute of renewable energy. It has abu7ndant supply. It is abundant in supply. Renewable sources of energy are also referred to as green technologies. They do not have an extensive environmental impact, unlike the fossil fuels. Renewable energy investments require high-cost investment.

The material and the labor cost are high rather than costly energy imports. There is a general increase in awareness about the environmental impact of burning fossil fuels. This has been possible largely due to advancement in mass communications due to progress in technology.

The significance of renewable sources of energy cannot be stressed enough. It is a clean source of energy and supports sustainable development. This has compelled all nations to realize the importance of these sources. All countries have invested in developing these resources and have engaged their scientist and engineers in this field.

4.3.1. Mitigating Climate Change

They are searching for new ways to utilize these sources of energy in an efficient way. Global warming is a serious environmental threat that is a consequence of the GHG. These gases are released by the burning of fossil fuels.

Fossil fuels have caused many unfortunate occurrences in the past. Only renewable sources of energy can offer a solution to mitigate the harmful impact of the global warming. They do not emit GHG while generating energy. This makes them the cleanest and the most viable option to mitigate climate change.

4.3.2. Infinite Supply

Fossil fuels are non-renewable and have a fixed supply. It is inevitable that these resources will exhaust one day. Therefore, before the crucial stage comes up, experts of energy sectors must maintain a positive attitude in this regard and should try their level best to replace fossils fuels with renewable energy sources as the main sources of generating electricity.

4.3.3. Competitive Pricing

Renewable energy sources are more reliable and abundant. Once the technologies are developed, and infrastructure is developed, it will become more cost-efficient. Besides, environmental impacts, renewable sources like the natural gas, oil, and petroleum have other drawbacks. The mining and drilling process involved in extracting these resources is hazardous.

The cost of extracting the same will further increase as these resources diminish further, and the population growth places further threat. Economies of scale and innovation are already resulting in renewable energies becoming the most sustainable solution, not only environmentally but also economically, for powering the world.

4.3.4. Minimize Environmental Impact

The GHG emitted by renewable energy is very low. This helps in mitigating the environmental threat caused by the combustion of fossil fuel.

The pollutants released when conventional sources are used to cause acid rain that erodes buildings and have a significant impact on the fragile ecosystem. They also deplete the ozone layer and cause global warming.

4.3.5. Creation of Job Opportunity

Renewable energy sources are still an emerging sector in most nations. Hence many companies would be inclined to invest in it. These companies will create many new job opportunities for the unemployed.

Hence it has a great potential to bring the unemployment rate down in many developing nations. This will have a significant positive impact on their economy.

4.3.6. Stability of Electricity Price

Renewable sources can stabilize the prices of electricity. This is because the only cost involved in electricity generated by renewable sources is the initial investment. Hence it does not fluctuate unlike the prices of coal, oil, and natural gases. The price of fuels is influenced by various occurrences around the globe, political being a chief cause. There are various instances in the past where political discord has led to serious crises. This is because there are few countries that are rich in these resources and are the major producers of the fuels.

Majority of the countries are dependent on these handfuls of countries. Renewable energy sources, on the other hand, are available to all countries. They can be locally produced and hence not influenced by political disturbances in distant countries.

4.3.7. Addresses Safety Concerns

There are various hazards associated with fossil fuel extraction. This includes explosion on oil platforms, collapsing of coal mines, etc. These safety concerns do not exist with renewable energy sources. The technology used in generating energy from renewable sources is not only eco-friendly, but is non-hazardous.

The conventional sources require the combustion of a massive quantity of natural gas in order to convert them into their usable form. It can also lead to calamitous accidents, an infamous incident being the British Petroleum Oil Spill in 2010. The renewable source can be harnesses with much less effort and are not difficult to find. Sources such as sunlight and wind energy are omnipresent.

The renewable sources of energy are safe and much cleaner. The exploration of petroleum and coal mining produces toxic wastes like lead, mercury, and other metals. The burning of coal to generate electricity uses large quantities of water which often discharges arsenic and lead compounds into surface waters and releases carbon dioxide, sulfur dioxide, nitrogen oxides, and mercury into the air. Similar problems arise with gasoline and similar products obtained from petroleum. These pollutants are a cause for much respiratory illness and can even be fatal.

4.3.8. Inexhaustible Source

The conventional sources of energy are finite. At one point in the distant future, these sources will cease to exist. On the other hand, clean energies are just as available as the sun from which they originate and adapt to natural cycles, hence their name "renewable." This makes it a tool of sustainable development. These sources can be used today without affecting the ability of future generation to meet their needs.

The conventional sources are veiled and restricted. The exact amount of these resources present underground or in the ocean bed is not known, but it is certain that they are finite and will exhaust one day. The use of these resources will diminish their quantity. Finding new sources will then become challenging and pricey. Even utilization of these resources will become risky and difficult.

4.3.9. Reducing Energy Dependence

The indigenous nature of clean sources gives local economies an advantage and brings meaning to the term "energy independence." When a nation is dependent on fossil fuel imports, it becomes subjugated to the political and economic short-term goals of the exporter country.

These can make the energy supply vulnerable. Whereas, the alternative sources are available everywhere throughout the world and reduces the dependency on other countries.

4.4. MARKET DEVELOPMENT FOR RENEWABLE ENERGY

The potential markets for renewable energy and the involvement of public sector is influenced by the particular conditions prevailing in the region. The use of green energy technologies and renewable energy will have huge benefits and support sustainable use of finite resources. The countries should attempt to strike a balance between liberalization and directing markets towards attaining the larger societal goals. The developing countries have to channelize more efforts to make the use of renewable energy sources viable.

Renewable energy is used both directly and indirectly. Direct use is the immediate use of renewable energy flows to satisfy energy service needs. Some of the renewable energy sources include solar heating, solar crop drying, and daylighting. There exist no energy markets. Policies can be formulated to encourage the use of renewable energy sources.

Indirect use of renewable energy refers to the generation of an energy carrier that is then applied in energy end-use equipment to provide the desired energy service. These energy carriers comprise of biogas, liquid biofuels, and electricity mechanical power. There are established markets for some of the carriers. In some cases, there is a more local use of these sources, for example, photo-voltaic for lighting and communication uses. In industrialized and transition countries and many developing countries, most renewable energy use takes place through markets for heat, electricity, and fuels. These kinds of markets are established in all emerging economies. Some have national wind systems for electricity and also an established market for fuel. Yet there are others that depend on direct uses of renewable energy and on markets established in the local areas.

In order to develop the markets for alternative sources, it is important to give battery incentives and establish regulations. Developing countries will benefit by finding new uses for renewable energy and focusing on direct end uses. The potential use for renewable energy varies from one setting to another. For instance, it would vary between urban and rural areas. Similarly, within urban areas, it would vary according to the level of development. There is thus not a single classification by national level of development, as most developing countries have segments of society that have well-developed energy markets in combination with large segments that have extremely poor access to energy services (Figure 4.4).

Figure 4.4: Renewable energy development in the California desert.

Source: https://farm9.staticflickr.com/8230/28902005814_a73990e694_b.jpg

4.4.1. Rural Energy

Accesses to sources are taken for granted by many privileged people who have unhindered access to these sources. There still 350 to 400 million household in the economically backward countries that have no electricity connections. This implies that there are no televisions and communication services. Traditional methods of illumination are used like fire and candles.

Renewable energy sources can be employed to address these gaps. There is a large potential to employ these resources, and the market for the same is in the developing stage.

- **Solar Home Systems:** They comprise of PV solar panel, charging controller, battery. It can also be used for other end uses like heating and lighting. One example is the solar lantern which can be used for illumination. Solar power market has been growing in, and large markets have been established. This trend is more prevalent in the rural areas of emerging economies. Installations may service single households or public buildings, like schools and health centers.
- **Biogas for Cooking and Lighting:** Biogas is produced with the help of digester which transforms the animal and plant wastes into fuels. These fuels are used for cooking, heating, lighting, and generating electricity. Digesters can be developed in different capacities as per the need.

Individual digester of small capacity can be built to meet the requirement of a single household, or larger ones can be used to cater to many households. There are various community and political issues and technical challenges that can hinder the market growth for these renewable sources.

- **Mini-Grids:** Small-grids can be useful for communities which have high population. The conventional source of energy for mini-grids is diesel. Alternative modes like solar PV, biomass digesters, and wind turbines can be used. They can either completely substitute diesel power or supplement it.
- **Small Scale Wind Power:** Wind power systems for individual families have been developed and tested in some nations. These systems have performed well, though some issues are experienced during the hotter months when the winds decrease. In order to combat this problem, many households have also installed solar PV to

4.4.2. Productive Uses in Rural Areas

In the rural areas, renewable energy sources find usage in agriculture, irrigation, and pumping water, cottage industries which uses sawmills and needs mechanical power. It can also be effectively employed in public services like health care and education sectors.

- **Agricultural Water Pumping:** Wind-powered water pumps have been used in rural areas for several decades since the historical times. In the recent times, other alternative sources like solar power PV driven water pumps, hybrid engines which run on dual fuels like biogas and diesel have also been developed and used.
- **Small Industry:** These can be driven by stand-alone energy systems. The development of these industries will create job opportunities and advance the economies. Mini-grids have a positive impact on development of industries. The mini-grids will become more economically feasible when there is a widespread adoption as a result of growing business. As the energy supply increases, new prospects come up, and the cycle of development continues.
- **Drinking Water:** Renewable energy sources work efficiently for purification of water. It can drive mechanical pump, filter the water, and even disinfect the water using UV and supply clean water. In emerging economies, it is gradually developing as a major market.
- **Crop Drying:** Crops are dried for extending their shelf life. Direct sun drying involves almost nil cost. Moreover, traditionally, it is considered as a healthier method compared to machine drying since it preserves the nutrients better. Particularly in the rural areas, direct solar radiation is used for drying crops.

4.4.3. Electrical Grids

Economic growth continues to drive electricity consumption, which raises concerns over how to best provide the power generation necessary for these rapidly expanding markets. The renewable sources are viable alternatives. Mini-hydro plant, geothermal power, biomass, wind farms are feasible options. Moreover, they have minimal environmental impact and helps in attaining social goals. This makes it an even more attractive option.

4.4.4. Cooking and Hot Water

Biomass is widely used in domestic and commercial cooking. It is directly combusted to heat food and water in emerging economies, particularly in the rural areas. However, forest resources have reduced in some nations, which make it necessary to find an alternative source of energy. Solar cookers can be cost-effective and efficient in cooking. However, more R&D is required to develop these technologies. Markets are primarily found where resource constraints are appearing.

Solar water heaters require no operating cost; however, it involves a onetime investment. It is still used widely but has to be developed, and it's used should be encouraged. The demand for domestic solar water heaters is growing worldwide.

4.4.5. Transport Fuels

Biomass can be used to generate liquid fuels. This can be used as transport fuels in various ways. Special engines can be designed that can run on ethanol; this increase the usage of ethanol as a fuel in its pure form. Ethanol can also be mixed with diesel or gasoline, which generates a substance called gasohol. This fuel is widely used in USA and Brazil to power vehicles. It reduces the consumption of diesel, which is a fossil fuel and has significant environmental impact.

Moreover, it is now proved that it is cost-effective to convert sugarcane to ethanol and use it power motor vehicles. The competitiveness of ethanol and gasohol relative to conventional gasoline has continued to improve, although the low price of oil and global automotive industry greatly affect the prospects of biomass-derived fuels, in the absence of accounting for extended costs and benefits.

4.5. THE RELATION BETWEEN RENEWABLE ENERGY AND SUSTAINABLE DEVELOPMENT

The relationship between RE and SD can be viewed as a hierarchy of goals and constraints that involve both global and regional or local considerations. The primary benefit is the mitigation of global warming, which is possibly the strongest incentive of renewable energy sources across the globe. Renewable energy sources do not release GHG. Hence it is important to develop technically feasible and cost-effective and reduce the economic burden.

Two important drivers here are technological advancement and optimal mitigation pathways. The technology, the investment involved, and the policy implementation is influenced by the community and the natural habitat they are a part of.

The issues of sustainability and their resolution are influenced by the following:

- Geographic setting. For instance, the amount of sunlight a place receives;
- Socio-economic situation like energy demand;
- The disparity within the social structures;
- Fragmented institutions;
- The condition of the present infrastructure.

A varying normative understanding of the connotation of sustainability (Lele and Norgaard, 1996).

Experts suggest that strategies and policy implementation based on the geographic settings and location (Wilbanks, 2002; Creutzig and Kammen, 2009) and a pluralism of epistemological and normative perspectives of sustainability (Sneddon et al., 2006).

These factors make it important to evaluate various impacts, and both environmental and social that arises due to the use of renewable energy sources. This is important to ensure that the renewable energy use is aligned with the overall sustainability goals.

There are some other considerations like these sources may also have some negative impact on the environment. It may deplete the natural resources. For instance, by up streaming GHG emission, it can impact the forests and affect the water bodies. Evaluating these impacts from the perspectives of the weak and strong sustainability paradigms elucidates potential tradeoffs between de-carbonization and other sustainability goals.

In order to align the efforts with sustainability goals can make consideration of other factors necessary. It may even compel the governments and the population at large to accept trade-offs. The additional constraints that are applied on energy systems can narrow down choices eliminating those which are not sustainable. Even if certain methods are technically viable, it may not meet the criteria for qualifying as sustainable development. However, the business-as-usual trajectories to which climate mitigation scenarios are compared are probably also insufficient to achieve SD (Sathaye et al., 2011).

4.5.1. Economic and Social Development and Human Well-Being

The energy sector plays a pivotal role in economic development. Energy consumption is positively correlated with economic advancement in majority of the countries all over the world. Economic development is measured by per capita GDP, and the human development index (HDI) is linked to per capita energy use.

The HDI used to compare levels of development among countries, includes as main items the purchasing power parity-adjusted income, the literacy and the life expectancy at birth. HDI is also an indicator of development level and societal well-being.

These two indicators are precise because data time series is obtainable. However, per capita, HDI and GDP are not precise measures of social and economic development. Another measure of technological development is the drop in energy intensity. Energy intensity is measured by the amount of energy consumed to produce one constant dollar of GDP.

The increase in employment rate is an added advantage which accrues with the use of renewable energy. The international community has also taken initiatives wherein financial commitments have been established to support the developing countries in mitigating climate change. The Copenhagen Agreements (CA, 2009) was aimed to attain this purpose. The outcome of the policies would depend on regional issues and the way it is implemented.

4.5.2. Renewable Energy, SD, and Energy Security

The energy security expression is highly context-dependent (Kruyt et al., 2009; Sathaye et al., 2011) and can best be interpreted as robustness against breakdowns of energy supply (Grubb et al., 2006). Energy security is different at different times and location. Two important aspects of energy security are the quantity of resources available and their dispersion and the predictability and dependability of energy source. They are relevant for the current systems as well as for planning the future systems for renewable energy.

It has been discussed that economic growth and energy consumption are correlated. Hence access to a reliable source of energy is economical and technical challenge and a major political preoccupation for emerging economies. This is because long breakdowns and functional problems are considered economic and functional hindrances (Larsen and Sonderberg,

2009). The fossil fuel is depleting, and the quality is also decreasing. This provides an incentive to migrate to renewable energy sources. Recoverable fossil fuel re-source amounts are controversial. Some experts have a positive view (Greene et al., 2006), while some have negative views (Campbell and Laherrere, 1998). Moreover, there are projections of lacking investments falling between both of them (IEA, 2009).

Using the renewable energy as an alternative helps in maintain the reserves of fossil fuels. This is because the fossil fuels are used at a lower rate and depleted less rapidly. Hence their exhaustion is postponed to the future and is further delayed.

The main concern of energy security is the availability and distribution of energy. There should be multiple sources of reliable energy supply. Dependence on a single source increases the vulnerability of energy system to disruption. Energy sources should be localized. Increased reliance on imports is also a potential source of vulnerability. This includes both the fossil fuels as well as the technology used.

Renewable sources are not concentrated in specific regions, unlike the fossil fuels. Moreover, they are not traded extensively in the world market. Hence it is imperative to develop renewable energy sources so that the reliance on actual energy imports is reduced (Grubb et al., 2006). Renewable energy diversifies the sources and increases the options of energy supply. This makes the economy more secure and curbs price volatility (Awerbuch and Sauter, 2006). This will help attain energy security at a local, national, and global level (Awerbuch and Sauter, 2006; Bazilian and Roques, 2008).

The introduction of renewable technologies adds new preoccupations to energy security. First, some unfriendly agents can obstruct energy supplies. Second, the dependence of energy supply on random and uncertain factors of nature like extreme drought. However, renewable energy can make energy supply more robust in rural areas that are remotely situated and have no access to grid. A wide range of energy sources coupled with good policy implementation and management can ensure energy security.

The indicators of energy security are difficult to identify. Some of the indicators are:

- The volume of reserves;
- The percentage of import in total primary energy consumption;
- The reserves-to-production ratio;
- The proportion of RE resources;

- The proportion of energy imports in total imports.

4.6. FUTURE OF RENEWABLE ENERGY SOURCES

Proper development of renewable energy sources is integral to maintain adequate supply for energy demand. Experts everywhere in the world are encouraging the use of renewable energy sources for generating power. Fuels prices are soaring, and power companies can no longer simply rely on fossil fuels to supply enough energy.

The renewable energy sources support sustainable development, and hence, power supply companies are more inclined to use them. Conservation of energy and optimum utilization of renewable energy is the ultimate goals to ensure sustainability. It is inevitable that the conventional sources of energy will deplete one day. Hence new sources of energy must be found to sustain the demand for energy.

More awareness has to be regarding conservation of energy. A simple example is ignoring lights left on. If lights are turned off when not in use, it results in considerable electricity savings. Compact Fluorescent Lamps (CFLs) can be used in place of incandescent lamps, which can give more lighting using the same amount of electricity. This results in savings as well as conservation of energy. Similarly using air conditioning systems and heating systems responsibly will also conserve energy.

Growth of green technologies for energy generation is indispensable. This has been supported by the statistics presented by the International Energy Agency (IEA) in 2015. They represented nearly half of all new electricity generation capacity installed in 2014, when they constituted the second biggest source of electricity worldwide, behind coal. As per the estimate of the IEA, by 2040, the demand for electricity will increase by 70%. The use of energy will also increase from 18 to 14% in the same time frame. This rise will be a consequence of advancements in emerging economies like China, India, Africa, South-East Asia, and Middle-East.

Clean energy sources are crucial for mitigating climate change and its destructive impact on the environment. Since the 19th century, the Earth's average temperature has increased by 0.85°C. The National Geographic brought attention to this fact in 2015 November. On the other hand, almost 17% of the global population, which is approximately 1.1 billion people still has no access to electricity. Moreover, 38% of the global population, which is approximately 2.7 billion people employs biomass for domestic purpose. This is hazardous to their health.

Considering the above situation, United Nations has set one of its objectives as ensuring access to electricity for all. The cut off year is 2030 to achieve this. Meanwhile, IEA that there will be acute shortage of electricity and 800 million people will be devoid of electricity if conventional sources of energy are relied on. This makes the target even more difficult to achieve.

The Paris Accord was entered into in 2015 December at the World Climate Summit. It ensures crucial support for renewable energy sources from the international community. The agreement, which will enter into force in 2020, establishes, for the first time in history, a binding global objective.

There were 200 signatory countries involved who vowed to control their emissions in order to ensure that there are no catastrophic climate change effects. The aim is to keep the climate change at 1.5°C much lower than the 2°C limit by the end of this century.

The adoption of green technologies based on renewable sources will also aid economic progress. The International Renewable Energy Agency (IRENA) projected that if the percentage of renewable energy is doubled by 2030 and taken to 36%, it will accelerate the global growth by 1.1% during that period. It will also result in increase in wellbeing of 3.7% and in employment in the sector of up to more than 24 million people, compared to 9.2 million today.

4.6.1. Renewable Energy Innovations

Innovations are a challenging task. There are various hurdles along the innovation chain. This includes the research and development phase; the pilot run to widespread diffusion. Some of these barriers reflect market imperfections; some inadequacies in the public sector domain; and some differences of view about needs, corporate priorities, relevant time horizons, and reasonable costs. Public support is needed to find a solution to these challenges. The amount of support needed would be determined by the maturity of the technology and its potential to capture the market. Hence, it will vary from one technology to another.

Direct government support is also crucial, especially for radically new technologies. Incremental advances are managed well by the private sector. The government support for a particular field of energy research ideally is determined by two factors. First, it is an area which is striving to adopt a sustainable energy policy and trying to transit from conventional modes. Second, to make national industries more competitive.

Another crucial factor is strong research to accomplish the objectives. Intervention is required whenever bottleneck is experienced in the innovation chain and specifically when the energy innovations hold great prospects. Increasingly, however, this chain is viewed as a complex, interactive system requiring networks of innovation, knowledge sharing, and demand 'pull' as well as supply 'push.'

The international community has become aware of the significance of renewable energy sources in the last two decades. Many countries have established new policies regarding targets setting, procurement of resources, green labeling, etc. Fiscal incentives have also been provided to encourage green technologies. Cost buy-down and widespread dissemination can be advanced through a number of policy measures.

An effective policy is the provision of temporary subsidies. This has been effective in Spain and Germany and has been instrumental in increasing the usage of renewable sources for generating energy. Many countries also establish markets for holder of green certificate. Temporary subsidiary is can be controlled better, and the outcome can also be predicted. It does not require any assumption about the price of the economic certificate.

Taxes have also been effective is making wide-spread adoption of renewable sources possible. Sweden is a good example, where carbon taxes have encouraged the use of biomass over coal in district heating systems. Brazil has also taken huge initiatives and established law to support the use of PV solar panels, wind energy, use of biomass, etc. The market alone cannot safeguard the interest of the nation and society, and hence, laws are needed.

4.7. CONCLUSION

Renewable energy is clean energy sources. They have an infinite supply, and with technical advancement, the cost for generating energy from renewable sources is becoming more competitive. They are considered superior to conventional energy sources due to many advantages they possess. There are various sources of renewable sources that helps diversify the energy portfolio. They are available in abundant quantity and can be used in any settings. The most significant benefit is that they do not release GHG and is an important step for mitigating climate change and reducing the pollutants released into the atmosphere.

On one hand, the cost for generating clean energy is reducing, and on the other hand, the cost of fossil fuels is rising with the depletion of these sources. This makes the use of these alternate sources inevitable. It can act as a substitute in countries which do not have fossil fuels, which reduces their dependency on imports and makes their energy system less vulnerable. Moreover, it delays the exhaustion of these scarce resources by reducing their consumption.

There are various types of renewable energy sources, namely solar energy, wind energy, geothermal energy, energy from oceans, hydroelectricity, and bio-energy. The last source is considered less clean than the other sources since it involves some emissions. Nevertheless, all these sources are important and can be harnessed with advancement in technology. The reduction of prices will ensure widespread adoption.

The most important aspect of renewable energy is that it is an important tool for attaining sustainable development goals, their primary function being ensuring judicious use of scarce resources, besides mitigating climate change and reducing emissions.

The international community has become aware of the significance of these technologies. Paris Accord of 2015 is an example of one such initiative at the global level. Many countries have taken concrete steps to mitigate to cleaner technologies by adopting various policies like tax imposition and providing incentives.

It is expected that the current efforts will continue and countries across the globe will remain committed to use the scarce resources judiciously and mitigate climate change. In order to achieve the said goals, renewable energy sources are indispensable.

REFERENCES

1. Acciona.com., (n.d.). *The Importance of Renewable Energy | ACCIONA*. [online] Available at: https://www.acciona.com/renewable-energy/ (Accessed on 21 May 2019).
2. Banton, C., (2019). *Why Renewable Resources are Becoming More Vital*. [online] Investopedia. Available at: https://www.investopedia.com/terms/r/renewable_resource.asp (Accessed on 21 May 2019).
3. Johansson, B. T., McCormick, K., Neij, L., & Turkenburg, W., (2004). *The Potentials of Renewable Energy*. [eBook] Available at: http://ren21.net/Portals/0/documents/irecs/renew2004/The%20Potentials%20of%20Renewable%20Energy.pdf (Accessed on 21 May 2019).
4. Marsh, J., (n.d.). *What are Examples of Renewable Energy Sources? | Energy Sage*. [online] Solar news. Available at: https://news.energysage.com/five-types-of-renewable-energy-sources/ (Accessed on 21 May 2019).
5. Ramos, P. M. N., (2013). *Renewable Energy and Sustainable Development*. [online] Redalyc.org. Available at: https://www.redalyc.org/html/301/30126353001/ (Accessed on 21 May 2019).
6. Shahzad, U., (2019). The need for renewable energy sources. [eBook] *ITEE Journal*. Available at: http://www.iteejournal.org/Download_August15_pdf_4.pdf (Accessed on 21 May 2019).
7. Sun power – United States, (2018). *Learn About Seven Types of Renewable Energy | SunPower Solar Blog*. [online] Available at: https://us.sunpower.com/blog/2018/02/23/learn-about-seven-types-renewable-energy (Accessed on 21 May 2019).

5

Electricity from Renewable Resources

CONTENTS

The chapter of electricity from the renewable resources talks about the generation or energy or electricity with the help of several numbers of power plants. This chapter provides highlights about the present scenario of the consumption of energy all across the world. This chapter emphasizes the various types of renewable sources which include solar energy and its types, hydro-electric power and its types, biomass energy and its types as well as the conversion of biomass into the energy, geothermal energy, and its types, ocean thermal and its types.

5.1. INTRODUCTION

Energy plays an essential part in the development and the growth of the economy for any nation or country. The utilization or consumption of the energy is increasing with a very huge rate all across the world. To satisfy the demand of the energy, various numbers of energy sources and effective use of the energy are being discovered. Several numbers of resources of renewable energy and the optimal use of the electricity are reviewed broadly.

Along with that, the inclination in the research and the development for the enhancement in the technologies of the consumption of energy and smart grid system for the future safety of energy are being explored. The outcomes show that the renewable resources for the creation of energy are becoming more widespread as more amount of the electricity production becomes an important necessity.

The production of the electricity could offer half of the total amount of the demand of the energy by the year 2050. To accomplish the demand of the energy in the future, the smart grid system can be used as an optimum and an effective system to the security of the energy. The smart grid also delivers the important benefits to the environment by conservation and renewable generation integration.

Also, the trend in research and development for the technological advancement of energy utilization and smart grid system for future energy security is presented. Results show that renewable energy resources are becoming more prevalent as more electricity generation becomes necessary and could provide half of the total energy demands by 2050. To satisfy the future energy demand, the smart grid system can be used as an efficient system for energy security. The smart grid also delivers significant environmental benefits by conservation and renewable generation integration.

The emission is hugely depending on the emission factor of the main sources of energy (that is input fuel of the plant). From all of the energy sources, the emission factor of the fossil fuels (that is coal, natural, gas, and oil) is very high. Fossil fuels are utilized on a huge demand as the main or primary fuel in the production of energy. In Malaysia, the fossil fuels (that is natural gas (near about 53%) and coal (near about 26%)) serve as a main or primary source in the production of electricity.

Wide range of the use of the fossil fuels, never the less, immensely impacts the environment. Based on the distribution of carbon dioxide (CO_2) in the year of 2013 all across the world. The breakdown of the emission is as follows: Coal (near about 43%), oil (near about 33%), gas (near about 18%), cement (near about 5%) and gas flare (near about 0.5%).

In the intervening time, renewable energy sources (such as solar, wind, hydro, geothermal, biomass, etc.) are those energy sources which are emission free in nature all across the world. Technologies for renewable energy are a superlative solution because they can participate in an important factor to the production of power with less emission of greenhouse gases (GHG) all across the globe. The "sustainable future" scenario of the International Energy Agency (IEA) shows near about 57% of electricity being offered by renewable energy sources by the year 2050, all across the globe.

A long-term prediction and planning are needed to accomplish this ultimate target. Production of electricity which is based on Renewable energy and supply to the national grid for a specific zone is important. The traditional grid sums the more than one network, and the system to control this contains several numbers of various degrees of communication and management, in which most of the systems are controlled with the help of manpower.

A smart grid is a fresh and new idea that takes in the directions to the innovation and transformation of the transmission and distribution grid. The smart grid system is the digital enhancement of the transmission and new markets for the various number of energy generation with the help of renewable energy sources. As per the current scenario, the smart grid is an often-cited term in the production of energy and distribution industry.

Smart grid linked with dispersed production of power is a new platform that essentially produces reliable security of supply (SOS) and quality of electric energy. This idea is based on the real world or practical and consistent as several numbers of energy sources become abundant, for example, solar, wind, biomass, and hydropower. Renewable energy sources, as well as the energy sources which are non-traditional in nature, are permitted to be merged with the distributed power generation link that has a smart grid.

5.2. WORLD ENERGY CONSUMPTION SCENARIO

Increment in the population all around the globe and enhancements in the standards of living can create a high demand of energy. This demand of energy all across the globe is increasing much rapidly as compared to the rate of the growth of the population. Near about 80% of the total primary energy is being supplied with the help of fossil fuels.

Consumption of energy all across the world from the year of 2002 to the year of 2030 indicates the increment in the demand of the energy by almost 60% (with an average of 1.5% every year). The demand of the energy will be near about 16.5 billion tons of oil equivalents (toe) by the year of 2030 as compared with near about 10.3 billion in the year of 2002.

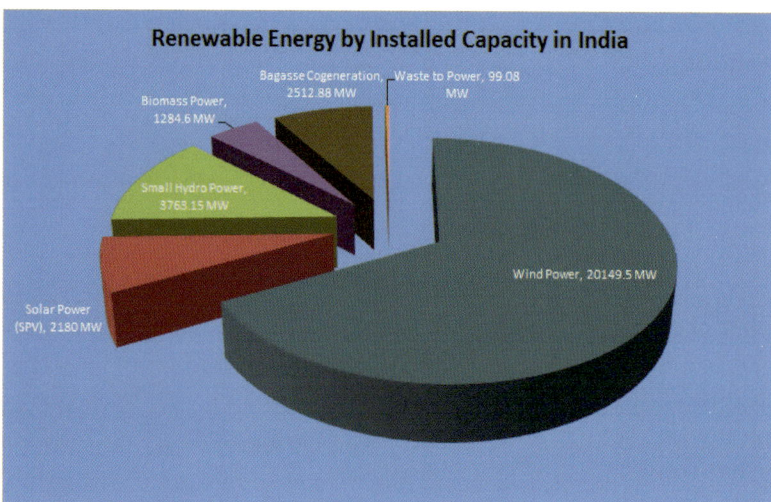

Figure 5.1: World energy consumption scenario.

Source: https://upload.wikimedia.org/wikipedia/commons/f/ff/Renewable_energy_share_India_2013.png

The primary energy consumption projection all across the world indicates that the fossil fuels will merely help in the biggest amount of energy, not taking into attention if its proportion will to some extent reduce from the value of 63% in the year of 2002 to 35% in the year of 2030. The demand for gas in the power sector will immensely elevate to a maximum value of 60% by the year of 2030, and the share of the market will elevate up to 47% by the year of 2030 as compared with the value of 36% in the year of 2002 (Figure 5.1).

In many of the regions of the world, natural gas is predicted to remain as a main inexpensive fuel in the new power station because of its efficiency is high. The share of coal in satisfying the total demand of primary energy will reduce from 30% in the year of 2012 to 20% by the year 2035.

The nuclear energy will reduce to 5% by the year 2030 from 7% in the year of 2002. The concentrations in various areas will remain to vary because of the degree of fluctuation in the economic growth, utilization of the energy of various users, prices of energy, geography, economic arrangement, culture, lifestyle, and climate.

5.3. TYPES OF RENEWABLE SOURCES

In the current situation, the united States count on heavily on coal, oil, and natural gas for its energy. Fossil fuels are non-renewable, that is, they draw on finite resources that will sooner or later decrease, becoming too costly or too environmentally damaging to retrieve. Dissimilarly, several numbers of renewable energy resources, for example, wind, and solar energy are continuously refilled or restocked and will never run out.

Most of the renewable energy comes either directly or indirectly from the sun. Sunlight, or solar energy, can be utilized directly for the heating and lighting homes as well as other buildings, for production of electricity, and for hot water heating, solar cooling, and a huge range of industrial and commercial uses.

The heat coming from the sun also drives the winds, and that energy is apprehended with wind turbines. Then, the winds and the heat coming from the sun become a source to evaporate the water. When this water vapor changes or converts into the rain or snow and flows downhill into the rivers or streams, its energy can be captured with the help of hydroelectric power. Simultaneously with the rain and snow, sunlight is the main reason in the growth of the plants. The organic matter that makes up those plants is known as biomass. Biomass can be utilized or applied to generate electricity, fuels for transportation, or chemicals. The utilization of the biomass for any of these reasons is called as bioenergy.

Hydrogen also can be found in many organic compounds, as well as water. Hydrogen is the element on the Earth which is widely available. But hydrogen does not occur naturally as a gas. Hydrogen always integrated with other various numbers of elements, for example, hydrogen integrated with oxygen to make water. Once separated from another element, hydrogen can be burned as a fuel or can be used to converted into electricity.

Not all of the renewable energy resources come from the sun. Geothermal energy taps the internal heat of the Earth for several numbers of uses, which also includes the production of electric power and the heating and cooling of buildings. And the energy of the tides of the Ocean comes from the pull of the gravity of the moon and the sun upon the Earth.

In the matter of fact, ocean energy comes from a several numbers of sources. Moreover, to tidal energy, there is the energy of the waves of the ocean, which are driven with the help of both the tides as well as the winds. The sun also helps to keep the surface of the ocean warm more than the depths of the ocean, producing a difference in the temperature that can be used as an energy source. All these forms of ocean energy can be used to generate the electricity.

5.4. SOLAR ENERGY

The sun does more than for planet Earth as compared to just provide light during the daytime. Every particle of sunlight (which is called photon) that reaches Earth, which carries energy that fuels the planet. Solar energy is the decisive source accountable for all of the weather systems, and energy sources exist on the Earth, and plenty of the solar radiation reaches to the surface of the planet with every hour to theoretically fill the requirements of energy for approximately a whole year, all around the world.

Where does all of this energy come from? Sun, which is similar to any other star present in the galaxy, is like a huge nuclear reactor. Deep in the core of the Sun, reactions of nuclear fusion generate huge amounts of energy that radiates in the outward direction from the surface of the Sun and into the space in the form of light and heat (Figure 5.2).

Figure 5.2: Solar energy.

Source: https://www.publicdomainpictures.net/pictures/270000/nahled/solar-energy-business.jpg

Solar power can be harnessed and converted to energy, which can be utilized with the help of photovoltaic (PV) or solar thermal collectors. Even though, solar energy only accounts for a small amount of overall use of energy all across the globe. The falling cost for the installation of solar panels means that more and more people in more places can take benefit of solar energy. Solar is a clean, renewable energy resource, and figures to act as a significant contribution in the future of the energy all across the globe.

5.4.1. Harnessing Solar Energy for Usable Power

There are several numbers of approaches to utilize the energy which is coming from the sun. The two prime approaches to utilize the energy from the sun are PV and solar thermal capture. PV are much more common to generate electricity in smaller scale (such as residential solar panel installations), and solar thermal capture is conventionally only used for the generation of electricity on huge scales in utility solar installations. Moreover, to the generation of electricity, lower temperature variations of solar thermal projects can be used for heating and cooling. Solar is one of the fastest rising and cost-effective sources of power which is used all across the world and will remain to spread very fast in the coming years. With the enhancements in the technologies of solar panel with each year, the economic advantages of solar expand, adding to the environmental perks of choosing a clean, renewable energy source.

5.4.2. Types of Solar Energy

Solar panels convert the light coming from the Sun into usable solar energy with the help of N-type and P-type semiconductor material. When the light from the Sun is captured by these semiconductor materials, the solar energy knocks electrons loose from their atoms, permitting the electrons to flow through the material to generate electricity.

This process of converting light (photons) to electricity (voltage) is called the PV effect. At present, solar panels convert most of the visible light spectrum and about half of the ultraviolet and infrared light spectrum to usable solar energy.

Technologies of solar energy utilize the energy of the Sun as well as light coming from the Sun to provide heat, light, hot water, electricity, and even cooling, for homes, businesses, and industry.

There are several numbers of technologies that have been developed to take benefits of solar energy.

5.4.2.1. Photovoltaic (Solar Cell) Systems

Solar cells convert the light coming from the sun directly into electricity. Solar cells are sometimes used as a power source for calculators and watches. Solar cells are made of a material which is semiconductor in nature similar to those used in computer chips. When the light coming from the Sun is captured by these materials, the solar energy knocks electrons loose from their atoms, permitting the electrons to flow through the material to generate the electricity. This process of converting light (which are a bundle of photons) to electricity (voltage) is called the PV effect.

Solar cells are usually integrated into modules that hold about forty cells. A number of these modules are mounted in PV arrays that can amount up to several meters on a side. These flat-plate PV arrays can be mounted at a fixed angle which is always facing south, or they can be mounted on a tracking device that follows the path of the Sun, permitting them to absorb most of the light coming from the Sun over the course of a day. Multiple numbers of connected PV arrays can offer a sufficient amount of power for a domestic purpose. For a large amount of electric utilization or industrial applications, hundreds of arrays can be organized and connected to form a single and a large PV system (Figure 5.3).

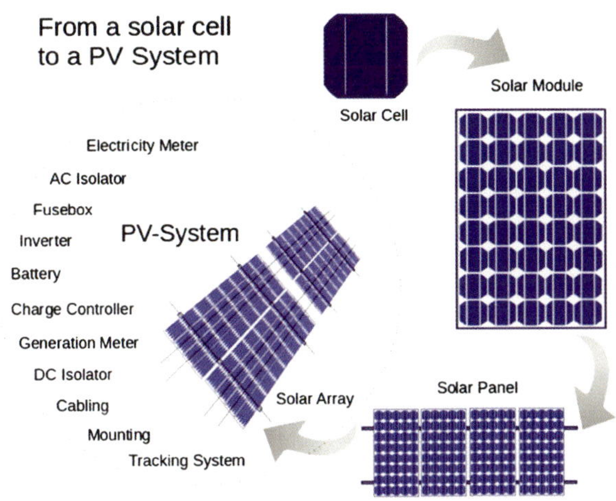

Figure 5.3: Photovoltaic (PV) system.

Source: https://commons.wikimedia.org/wiki/File:From_a_solar_cell_to_a_PV_system.svg

Thin film solar cells use layers of semiconductor materials only a few micrometers thick. Thin film technology has made it practically possible for solar cells to now two times as rooftop shingles, roof tiles, building facades, or the glazing for skylights or atria. The solar cell version of items such as shingles offers the same protection and durability as ordinary asphalt shingles.

Some of the solar cells are designed to operate with focused sunlight. These solar cells are built into concentrating collectors that use a lens which helps to focus the sunlight onto the cells. This method has both advantages as well as disadvantages as compared with the flat-plate PV arrays. The main concept is to utilize very little of the expensive semiconducting PV material even though collecting as much light coming from the sun as possible.

But because the lenses must be pointed at the sun, the use of focused collectors is restricted to the sunniest region of the nation. Some focused collectors are constructed in such a way that they are mounted on simple tracking devices, but most require sophisticated tracking devices, which in future restrict their use to electric utilities, industries, and large buildings. The execution or implementation of a solar cell is evaluated in terms of its effectiveness and productivity at converting light coming from the sun into electricity. Only light from the sun of certain energies will work proficiently to generate electricity, and much of it is reflected or captured by the material that makes up the cell. Because of this, a conventional commercial solar cell has an effectiveness of near about 15% or about one-sixth of the light coming from the sun striking the cell produces electricity.

Low effectiveness means that larger PV arrays are required, and that means higher cost. Enhancement in the effectiveness of the solar cell while holding down the price per cell is a crucial objective of the PV industry, researchers from National Renewable Energy Laboratory (NREL), and other United States Department of Energy (DOE) laboratories, and they have made an important progress. The first solar cells, built in the time period of 1950s, and these solar cells had efficiencies of less than 4%.

5.4.2.2. Solar Hot Water

The shallow water of a lake is commonly much warmer as compared to the deep water. That is because the light coming from the sun can heat the lake bottom in the shallow areas, which in turn, heat up the water. It's way of nature of solar water heating. The heat coming from the sun can be utilized in basically with the same approach to heat water, which is used in buildings and swimming pools.

Most of the heating systems based on solar water for buildings have two prime elements, which are a solar collector and a storage tank. The most general collector is known as a flat-plate collector. A flat-plate collector is mounted on the roof; it is consisting of a thin, flat, rectangular box with a transparent cover that faces the sun.

Small sized tubes run through the box and carry the fluid which either water or other fluid, for example, an antifreeze solution which is needed to be heated. The tubes are connected to an absorber plate, which is painted in black color so that it can absorb more amount of the heat. As heat builds up in the collector, it heats the fluid which is flowing through the tubes (Figure 5.4).

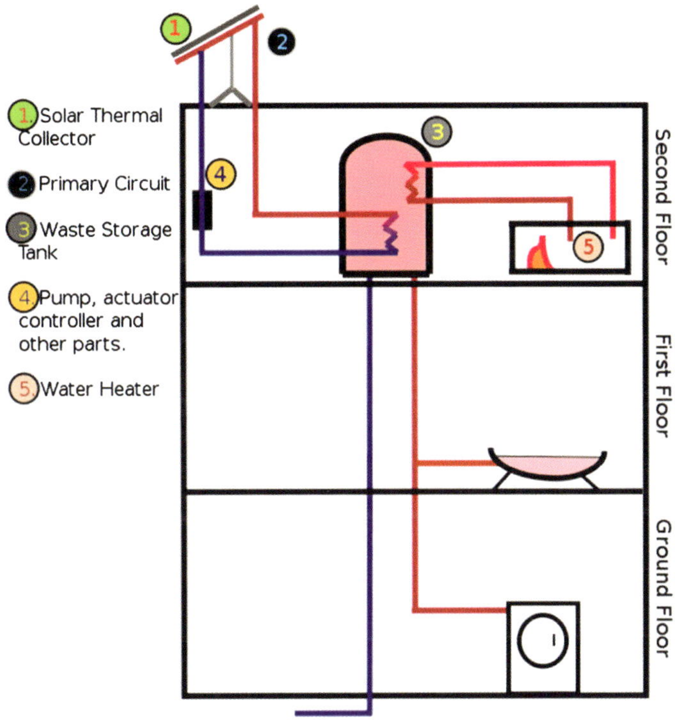

Figure 5.4: Solar hot water system.

Source: https://en.wikipedia.org/wiki/File:Solar_Water_Heater_Diagram.svg

After then, the storage tank holds the hot liquid. It can be just a modified water heater, but it is commonly larger and very well insulated. Systems that use fluids other than water generally heat the water by flowing it through a coil of tubing in the tank, which is full of hot fluid.

Solar water heating systems can be either active or passive. But the most usually used heating systems are active systems. Active systems depend on the pumps, which help to move the liquid between the collector and the storage tank. On the other hand, passive systems hold on gravity and the tendency for water to naturally circulate as it is heated.

Swimming pool systems are much simpler. The filter pump of the pool is used to pump the water through a solar collector, which is commonly made of black plastic or rubber. And of course, the pool stores the hot water.

5.4.2.3. Solar Electricity

In the present time, many of the power plants utilize fossil fuels as a heat source to boil the water. The steam coming from the boiling water helps to rotate a large turbine, which drives a generator that in turn, generate electricity.

Never the less, a new generation of power plants, with focused solar power systems, utilizes the heat coming from the sun as a heat source. There are three most important and prime factor of focused solar power systems, which are parabolic-trough, dish or engine, and power tower (Figure 5.5).

Figure 5.5: Installation of solar panels.

Source: https://cdn.pixabay.com/photo/2016/11/03/15/03/solar-panels-1794467_960_720.jpg

Parabolic-trough systems converge the energy coming from the sun with the help of long rectangular, curved (U-shaped) mirrors. The mirrors are inclined in the direction of the sun, which help in focusing the light coming from the sun on a pipe that drives down the center of the trough. This heats the oil passing inside the pipe. Then the hot oil is utilized to boil the water in a traditional steam generator to generate electricity.

A dish system or engine system uses a mirrored dish (just same to a very large satellite dish). The dish-shaped surface collects and converges the heat coming from the sun onto a receiver, which captures the heat and transfers it to fluid within the engine. The heat helps the fluid to expand against a piston or turbine to generate mechanical power. Then the generated mechanical power is utilized to drive a generator or alternator to generate electricity.

A power tower system uses a large field of mirrors to focus or converge the light coming from the sun onto the top of a tower, where a receiver is placed. This heat molten salt passing with the help of the receiver. Then, the heat of the salt is utilized to produce the electricity with the help of a typical steam generator. Molten salt retains heat effectively, so it can be stored for longer time before being transformed into the electricity. That means electricity can be generated on the cloudy days or even at that time when several hours after sunset.

5.5. HYDRO-ELECTRIC POWER

Flowing water generates energy that can be absorbed and converted or transformed into electricity. The process of conversion is called hydroelectric power or hydropower.

The most usual type of hydroelectric power plant utilizes a dam on a river to hold the water in a reservoir. Water released from the dam or reservoir flows with the help of a turbine, the water drives the turbine, which in turn drives a generator to generate electricity.

But hydroelectric power does not importantly need a large dam. Some of the hydroelectric power plants just use a small-sized canal to give the direction of the flow of the water from the river water with the help of a turbine.

Another type of hydroelectric power plant which is called a pumped storage plant. A pump storage plant can even store power. The power is sent from a power grid into the electric generators. The generators then drive the turbines in backward direction, which helps the turbines to pump the water from a river or lower reservoir to a reservoir which is situated at higher place, where the power is stored. To utilize the power, the water is freed from the upper reservoir back down into the river or lower reservoir. This drives the turbines in forward direction, activating the generators to generate electricity.

A small or micro-hydroelectric power system can generate sufficient amount of the electricity for a home, farm, or ranch (Figure 5.6).

Figure 5.6: Nizhne-Bureiskaya hydroelectric power plant.

Source: http://static.kremlin.ru/media/events/photos/big2x/eVaJmUYuqssqz7n-zX2TbZqkRNbftJfAv.jpg

5.5.1. Types of Hydro-Electric Power

Hydropower is power derived from the force which is generated from the movement or flow of the water. It is broadly utilized to generate electricity between the other useful reasons.

Hydropower is a versatile, flexible technology that at its smallest can power a single home, and at its largest can supply the power to an entire industry and the public with renewable electricity on a national and even regional scale. In terms of potential of generation of electricity, hydro accounts for eight of the ten biggest power stations all across the world.

There are four main types of hydropower that are used to generate electricity broadly:

5.5.1.1. Run-of-River Hydropower

Run-of-river hydropower is a facility that provides the direction to the flowing water from a river with the help of a canal or penstock to drive or rotate a turbine. Usually, a run-of-river project will have little or no facility to store electricity.

Run-of-river offers a non-stop supply of electricity (which is baseload), with some flexibility of operation for daily variations in demand with the help of water flow that is controlled by the facility (Figure 5.7).

Figure 5.7: Run-of-river hydropower.

Source: https://cdn.pixabay.com/photo/2017/10/06/16/21/hydroelectric-power-station-2823691_960_720.jpg

5.5.1.2. Storage Hydropower

Usually, a large system that utilizes a dam to store water in a reservoir. Electricity is generated by discharging the water from the reservoir with the help of a turbine, which spins and activates a generator.

Storage hydropower offers baseload as well as the potential to be switch off and started up at short notice depending on the demands of the system (which is peak load). Storage hydropower can provide sufficient amount of storage potential to operate self-sufficiently of the hydrological inflow for many weeks or even months (Figure 5.8).

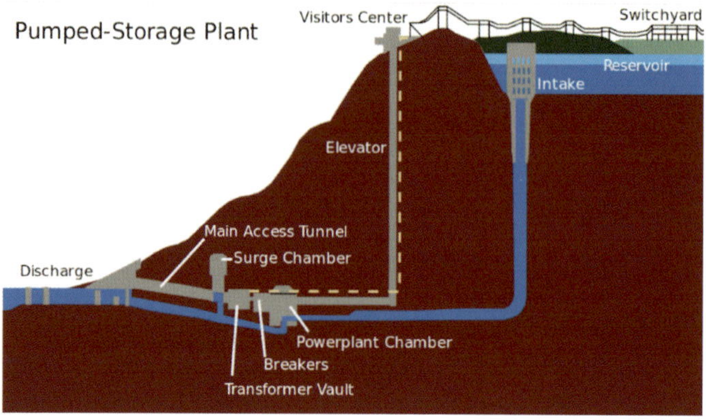

Figure 5.8: Storage hydropower.

Source: https://upload.wikimedia.org/wikipedia/commons/thumb/8/89/Raccoon_Mountain_Pumped-Storage_Plant.svg/1280px-Raccoon_Mountain_Pumped-Storage_Plant.svg.png

5.5.1.3. Pumped-Storage Hydropower

Pumped-storage Hydropower offers a supply of peak-load, connecting water which is cycled between a lower and upper reservoir with the help of pumps which utilizes the surplus energy from the system at times of low demand. When the demand of the electricity is high, water is discharged back to the lower reservoir with the help of turbines to generate the electricity (Figure 5.9).

Figure 5.9: Pumped-storage hydropower.

Source: https://commons.wikimedia.org/wiki/File:Ming-Tan_Power_Station_Pumped_Storage_Power_Plant09.jpg

5.5.1.4. Offshore Hydropower

Offshore Hydropower a less recognized but emerging group of technologies that utilizes the tidal currents or the power of waves to produce electricity from seawater (Figure 5.10).

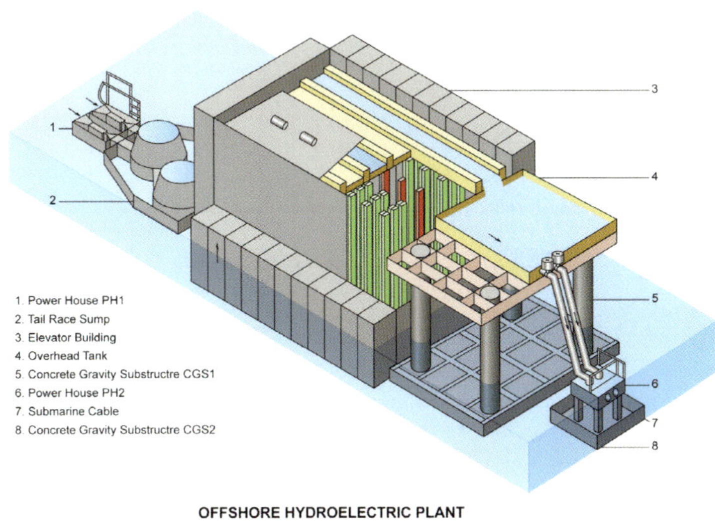

Figure 5.10: Offshore hydropower.

Source: https://upload.wikimedia.org/wikipedia/commons/d/d1/OFFSHORE_HYDROELECTRIC_PLANT_DESIGN_CONFIGURATION.jpg

5.6. BIOMASS ENERGY

Biomass is an organic material that comes from the waste of the plants as well as animals. Biomass is a renewable source of energy. Biomass encompasses the stored energy from the sun. Plants capture the energy coming from the sun, and that process is called photosynthesis. When biomass is scorched, the chemical energy in biomass is freed or discharged in the form of heat. Biomass can be scorched directly or converted to liquid biofuels or biogas that can be utilized in the form of fuels (Figure 5.11).

Figure 5.11: Biomass boiler at croft castle, Herefordshire.

Source: https://farm2.staticflickr.com/1625/24290854501_4c294d3945_b.jpg

5.6.1. Examples of Biomass and Their Uses for Energy

Wood and wood processing waste which is scorched to increase the temperature inside the buildings, to generate process heat in the industry, and in the production of electricity. Agricultural crops and waste materials which is burned in the form of a fuel or transformed into liquid biofuels. Food, yard, and wood waste in garbage which is scorched for the production of electricity in power plants or transformed into biogas in landfills. Animal manure and human sewage which is converted or transformed into biogas, which can be burned in the form of a fuel.

Biomass power is carbon neutral electricity produced with the help of renewable organic waste that would else be dumped in landfills, burned in open air, or left as fodder for forest fires.

When these materials are burned, the energy in biomass is freed is in the form of heat. If someone has a fireplace that person already is participating in the utilization of biomass in the form of the wood that individual burn in it is a biomass fuel.

In biomass power plants, wood waste or other waste is burned to generate the steam that helps to spin or drive a turbine to generate electricity, or that offers heat to the industries as well as homes. Providentially, new technologies which include pollution controls and combustion engineering which have enhanced to the point that any discharges from the burning of biomass in industrial facilities are commonly less than emissions generated when the utilization of fossil fuels (coal, natural gas, oil). Re-Energy has involved these technologies in some facilities.

5.6.2. Biomass Challenges

Through the course of the process to generate or produce electricity is same whether the utilization of a biomass fuel or a fossil fuel, the gear required inside the plant is dissimilar. All of the power generation with the help of re-energy facilities has been outfitted, and new procurements are upgraded, which helps to allow for the burning of biomass.

As with any process of the generation of electricity, the facility requires a constant amount of supply of fuel. In all cases, re-energy has suppliers to deliver a constant stream of biomass and has engaged other suppliers to make sure the facilities have what they require.

There are two issues that impacts or influence the assessment of the biomass as a worthwhile solution to issues related to energy. These issues are the effects of the farming and generation of biomass and the effects of the factory conversion of biomass into usable energy or electricity. There are as many environmental and economic advantages as there are disadvantages to every problem, which shows a difficult task in assessing the capable success of biomass as an alternative fuel.

For example, the substitute of coal by biomass could outcome in "a considerable reduction in net carbon dioxide emissions that contribute to the greenhouse effect." On the other hand, the utilization of wood and other plant material for fuel may cause the deforestation. In the present time, people are all attentive of the issues linked with stripping off the forests, and extensive clear-cutting can lead to groundwater contamination and irreversible erosion patterns that could literally alter the structure of the ecology all across the world.

Biomass must be well thought-out in the exploration for a substitute source of energy which is available in a huge quantity yet non-disruptive means, and in the meantime, it has the potential of being executed at all degrees of society. Even though tree plantations have "considerable promise" in supplying an energy source, "actual commercial use of plantation-grown fuels for power generation is limited to a few isolated experiences."

5.6.3. Converting Biomass to Energy

Solid biomass, for example, wood and garbage, can be scorched directly to generate heat. Biomass can also be transformed into a gas which is called as biogas or into liquid biofuels, for example, ethanol and biodiesel. These fuels can then be scorched to produce the energy, and these energies can be utilized anywhere.

Biogas is generated when paper, food scraps, and yard waste decompose in landfills, and it can be generated with the help of processing sewage and animal manure in distinct vessels, which is known as digesters (Figure 5.12).

Figure 5.12: Drax power station for the conversion of biomass into electricity.

Source: https://s0.geograph.org.uk/geophotos/03/58/43/3584383_1487 6a43.jpg

Ethanol is made from crops, for example, corn and sugar cane that are agitated to generate fuel ethanol and can be utilized to power up the vehicles. Biodiesel is generated with the help of vegetable oils and animal fats and can be utilized in the vehicles and as heating oil.

Biomass fuels offer near about 5% of the total primary energy use in the United States in the year of 2017. Of that 5%, near about 47% was created with the help of biofuels (mainly ethanol), 44% was generated with the help of wood and wood-derived biomass, and 10% was generated with the help of the biomass in municipal waste (sum of all percentages is greater than a 100% because of independent rounding).

Researchers are attempting to create a number of approaches other than this to utilize more of biomass for fuel and electricity generation.

5.6.4. Types of Biomass

5.6.4.1. Wood and Agricultural Products

Most biomass used today is homegrown energy. Woods components such as logs, chips, bark, and sawdust are accounts for near about 44% of biomass energy. But any organic matter can generate biomass energy. Other biomass sources can contain agricultural waste products such as fruit pits and corncobs. Wood and wood waste are being used to produce energy or electricity (Figure 5.13).

Figure 5.13: Wood and agricultural products.

Source: https://pxhere.com/en/photo/1192653

Much of the electricity is utilized by the industries making the waste, it is not distributed by utilities, and it is a process which is known as cogeneration. Paper mills and sawmills utilize much of their waste products to produce steam and electricity for their own use. Never the less, meanwhile they utilize so much amount of energy, they require to buy an extra energy or electricity from the utilities.

5.6.4.2. Solid Waste

Scorching trash turns waste into a usable form of energy. One ton (near about 2,000 pounds) of garbage comprises near about as much heat energy as 500 pounds of coal. Garbage is not all biomass, but it could be half of its energy content comes from plastics, which are created with the help of petroleum and natural gas.

Power plants that scorch the garbage for the production of energy are called waste-to-energy plants. These plants produce electricity much as coal-red plants do, apart from that combustible garbage is not coal but is the fuel which is utilized to heat their boilers (Figure 5.14).

Figure 5.14: Solid waste.

Source: https://cdn.pixabay.com/photo/2017/06/25/18/06/scrap-yard-2441432_960_720.jpg

5.6.4.3. Landfill Gas and Biogas

Bacteria and fungi are not picky eaters. They eat dead plants and animals, triggering them to rot or decay. A fungus on a rotting log is converting cellulose to sugars to feed itself. Even though this process is decelerated in a landfill, a constituent which is called as methane gas is still generating as the waste decays. New regulations need landfills to gather the content of the methane gas for safety and environmental reasons (Figure 5.15).

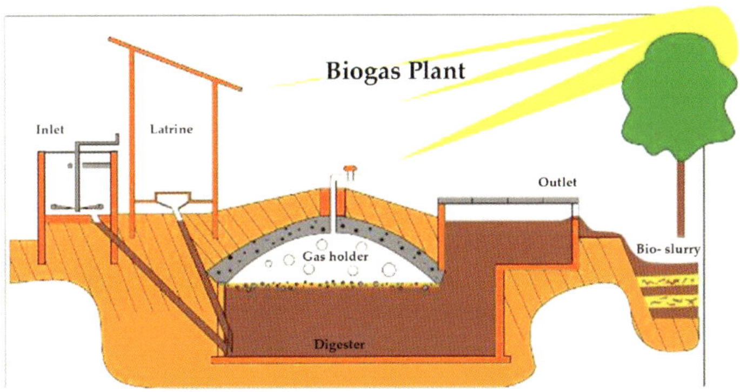

Figure 5.15: Landfill gas and biogas.

Source: https://upload.wikimedia.org/wikipedia/commons/0/02/Biogas_plant_sketch.jpg

Methane gas is colorless as well as odorless in nature, but methane is not harmless. The gas can be the reason for explosions if it seeps into nearby homes and is ignited. Landfills can gather the methane gas, cleanse the methane gas, and utilizes it in the form of fuel. Methane can also be generated with the help of energy from agricultural and human wastes.

Biogas digesters are airtight vessels or pits lined with steel or bricks. Waste put into the vessels is fermented without oxygen to generate the gas which is enriched with the content of methane. This gas can be utilized to generate the electricity, or for cooking and lighting purposes.

5.6.4.4. Ethanol

Ethanol is an alcohol fuel (also known as ethyl alcohol) created with the help of process of fermenting the sugars and starches found in plants and then distilling them. Any organic material which is consisting of cellulose, starch, or sugar can be made into ethanol. Most of the ethanol generated in the United States comes from corn. New techniques are generating ethanol with the help of cellulose in woody fibers from trees, grasses, and crop residues.

In the present time, almost all of the gasoline sold in the United States consist near about 10% of ethanol and is known as E10. In the year of 2011, the United States Environmental Protection Agency (EPA) accepted the introduction of E15 (15% ethanol and 85% of gasoline) as the application in the passenger vehicles from model year of 2001 and later (Figure 5.16).

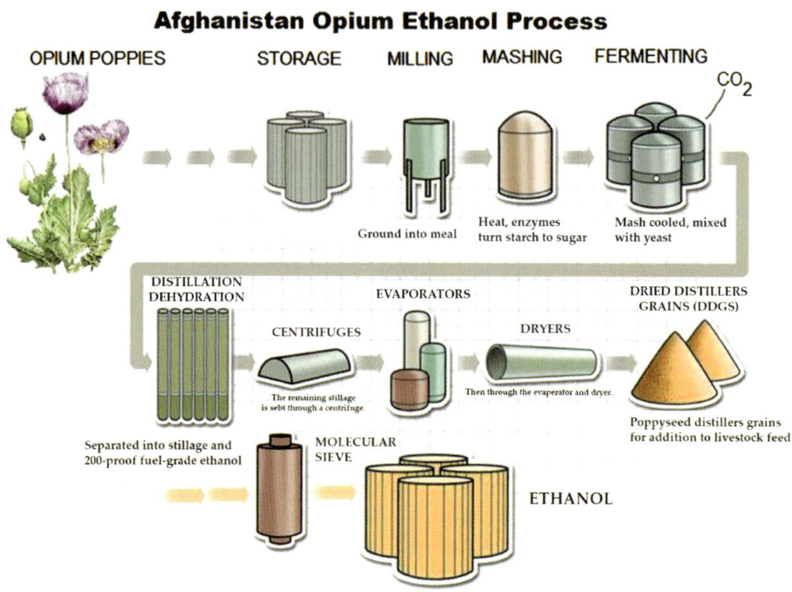

Figure 5.16: Afghanistan opium ethanol process.
Source: https://upload.wikimedia.org/wikipedia/commons/4/4f/Afghanistan_Opium_Ethanol_Energy_-_04082008.jpg

Fuel, which consists of 85% ethanol and 15% gasoline (E85) qualifies as a substitute to the fuel. There are more than ten million flexible fuel vehicles (FFV) on the road that can run with the effectiveness on E85 or E10, just under 10% of these vehicles use E85 on a regular basis.

5.6.4.5. Biodiesel

Biodiesel is a fuel which is generated with the help of chemically reacting alcohol with the vegetable oils, animal fats, or greases, for example, recycled restaurant grease. In the present time, the majority of biodiesel is generated with the help of soybean oil. Biodiesel is sometimes also blended with petroleum diesel in ratios of 2% (B2), 5% (B5), or 20% (B20).

Biodiesel can also be utilized as neat (pure) biodiesel (B100). Biodiesel fuels are companionable with and can be utilized in diesel engines which are not modified with the existing fueling infrastructure (Figure 5.17).

Figure 5.17: Argent energy biodiesel plant.

Source: https://upload.wikimedia.org/wikipedia/commons/1/18/Argent_Energy_Biodiesel_Plant_-_geograph.org.uk_-_232328.jpg

It is one of the quickly increasing or emerging transportation fuels in the United States Biodiesels are consists virtually no sulfur, so it can decrease the level of sulfur in the diesel fuel supply all across the nation, even compared with low sulfur fuels which is used in present time.

Despite the fact that, taking away sulfur from diesel which are based on petroleum outcomes in poor lubrication, biodiesel is a main or prime lubricant and have the potential to decrease the friction of diesel fuel in blends of only 1 or 2%. This is a significantly crucial characteristic because the EPA now needs that sulfur levels in diesel fuel be 97% lower than they were prior to the year of 2006.

5.7. GEO-THERMAL ENERGY

Geothermal energy is basically the heat emitted from the Earth. Geothermal energy is clean and sustainable in nature. Resources of the Geothermal energy has a wide range from the shallow ground to the hot water and hot rock found a few miles underneath the surface of the Earth, and down even deeper to the immensely high temperatures of molten rock which is known as Magma (Figure 5.18).

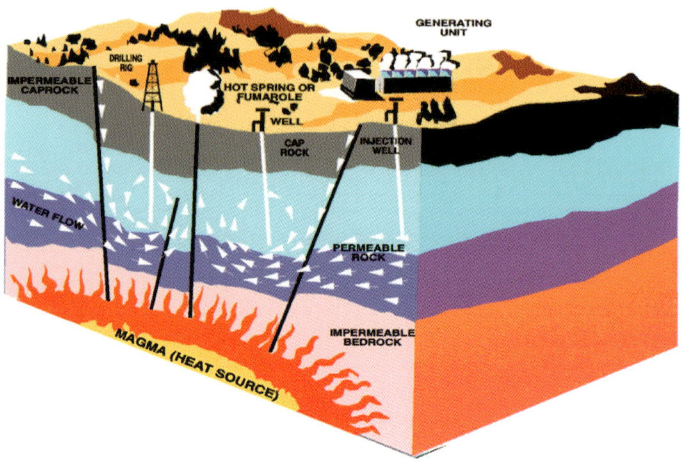

Figure 5.18: Geothermal energy conversion.

Source: https://upload.wikimedia.org/wikipedia/commons/c/c4/Geothermal_energy_methods.png

Nearly at every place, the shallow ground or ten feet upper of the surface of the Earth keeps an almost continuous tor constant temperature between the range of 50°F and 60°F (which is equal to10°C and 16°C). Geothermal heat pumps can tap into this resource for heating as well as cooling the buildings.

A pump system based on geothermal heat contains a heat pump, an air delivery system (called as ductwork), and a heat exchanger (a network of pipes buried in the shallow ground near the building). In the time of winter, the heat pump eliminates the heat coming out from the heat exchanger and transmits with the help of the pump into the indoor air delivery system. In the time of summer, the process is upturned, and the heat pump uses to transmit the heat from the indoor air into the heat exchanger. The heat removed from the indoor air during the time of summer can also be utilized to offer a free source of hot water.

In the United States, most geothermal reservoirs of hot water are situated in western states such as Alaska and Hawaii. Wells can be bored into underground reservoirs for the production of electricity. Some of the geothermal power plants utilize the steam from a reservoir to provide power to a turbine or a generator, on the other hand, others utilize the hot water to boil a working fluid that turns to vapor and then drives a turbine.

Hot water near the surface of Earth can be used directly for heat. Direct-use applications which involve heating buildings, growing plants in greenhouses, drying crops, heating water at fish farms, and several numbers of industrial processes such as pasteurizing process of milk.

Hot, dry rock resources occur at depths of three to five miles everywhere underneath the surface of the Earth and at lesser depths in specific areas. Access to these resources which includes injecting cold water down one well, flowing it with the help of hot fractured rock, and drawing off the heated water from another well. In the present time, there are no commercial applications of this technology. The present technology also does not yet permit the recovery of heat directly from magma, the very deep and most powerful resource of geothermal energy.

Most of the power plants require steam to produce the electricity. The steam helps to drive the turbine that triggers the generator, which in return generates the electricity. Many of the power plants still utilize the fossil fuels to boil water to produce steam. Geothermal power plants, nevertheless, utilizes the steam which is produced from reservoirs of hot water found a couple of miles or more below the surface of the Earth. There are mainly three types of geothermal power plants which are dry steam power plant, flash steam power plant, and binary cycle power plant.

5.7.1. Dry Steam Power Plants

Dry steam power plants draw from beneath the ground resources of steam. The steam is transmitted with the help of pipes directly from underground wells to the power plant, where it is focused into a turbine or generator unit.

There are only two known underground resources of steam in the United States: The Geysers in northern California and Yellowstone National Park in Wyoming, where there is a well-known geyser which is known as Old Faithful. Through the course of time, Yellowstone is safeguarded from development, the only dry steam plants in the country are at The Geysers.

5.7.2. Flash Steam Power Plants

Flash steam power plants are the most commonly used power plant which produces electricity. Flash Steam power plant utilizes the geothermal storages of water having the temperatures greater than 360°F or near about 182°C.

This very hot water is transmitted up with the help of wells in the ground under its own pressure. As it transmitted in the upward direction, the pressure reduces, and some of the hot water boils converts into the steam. The produced steam is then taken out from the water and then applied to provide as a power source for a turbine or a generator. Any leftover water and condensed steam are transmitted back into the reservoir, making this a sustainable resource.

5.7.3. Binary Cycle Power Plant

Binary cycle power plants functions on water at lower temperatures of near about 225° to 360°F or approximately 107°–182°C. Binary cycle power plants utilize the heat from the hot water to boil a working fluid, typically an organic compound having a low boiling point.

The working fluid is converted into a vapor state with the help of heat exchanger and utilizes to drive the turbine. The water is then injected back into the ground again to be heated. The water and the working fluid are set aside separated through the course of the whole process, so there are little or no air emissions.

Small-scale geothermal power plants (within the five megawatts) have the capability for the extensive application in rural areas, perhaps even as distributed energy resources (DERs). DERs mention to a several numbers of small, technologies for the modular power-generating that can be integrated to enhance the functioning of the electricity delivery system.

5.8. OCEAN ENERGY

The ocean can generate two types of energies, which are thermal energy with the help of the heat coming from the Sun and mechanical energy with the help of the tides and the waves.

Oceans conceal more than 70% of the surface of the Earth, making them the largest solar collectors all across the planet. The heat coming from the Sun helps to warm the surface of the water a lot more as compared to the deep ocean water, and this difference in the temperature produces thermal energy. Just a small share of the heat that is captured in the ocean could provide power all across the world.

Ocean thermal energy can be utilized for several numbers of applications, which also includes the generation of electricity. There are primarily three types of electricity conversion systems, named as closed-cycle, open-cycle, and hybrid (Figure 5.19).

Figure 5.19: OTEC in Hawaii.

Source: https://upload.wikimedia.org/wikipedia/commons/3/33/OTEC_in_Hawaii.jpg

Closed-cycle systems utilize the warm water from the surface of the ocean to convert into the vapor state as working fluid, which has a low-boiling point, for example, ammonia. The vapor expands and helps to drive the turbine. The turbine then triggers the generator for the generation of electricity.

Open-cycle systems actually boil the seawater by operating at low pressures. This generates steam that flows through a turbine or generator. And hybrid systems combine both closed-cycles as well as open-cycle systems.

Ocean mechanical energy is relatively dissimilar as compared to the ocean thermal energy. Although the sun impacts all activities occur in the ocean, tides are driven primarily with the help of the gravitational pull occur due to the moon, and waves are driven primarily with the help of the winds. As an outcome, tides, and waves are alternating sources of energy; on the other hand, the ocean thermal energy is equally continuous. Also, dissimilar as compared to the thermal energy, the conversion of the electricity of both tidal as well as wave energy typically involves mechanical devices.

A barrage (dam) is commonly used to convert tidal energy into the electricity with the help of high-speed water striking the blades of turbine with the help of nozzles, which in return triggering the generator. For the conversion of wave energy, there are three basic systems, which are channel systems that transmit the waves into reservoirs.

Second is float systems that helps to runs the hydraulic pumps. And last but not the least oscillating water column systems that use the waves to compress air within a container. The mechanical power produces with the help of these systems either directly triggers the generator or transmit to a working fluid, water, or air, which then runs the turbine or generator.

5.8.1. Offshore Wind Energy

Wind energy is an indirect form of solar energy. Wind energy is valued that near about 1–2% of the solar radiation that penetrates the earth is then transformed in to wind energy. Generally, wind outcomes from a non-uniform heating of various regions of the earth, causing cooler, dense air to circulate to substitute warmer light air.

On the other hand, some of the energy coming from the Sun is captured directly by the air; most of the energy in the wind is first captured by the surface of the earth and then transmitted to the air with the help of convection (Figure 5.20).

Figure 5.20: Offshore wind energy.

Source: https://upload.wikimedia.org/wikipedia/commons/5/54/Sheringham_Shoal_Wind_Farm_2012.jpg

5.8.2. Ocean Current Energy

Regions that usually experience high marine current flows are in narrow straits, between the islands, and around headlands.

Entrances to lochs, bays, and large harbors sometimes also have high marine current flows. Typically, the resource is largest where the water depth is comparatively empty or shallow, and a good tidal range exists. Specifically, large marine current flows exist where there is a considerable difference in phase between the tides that flow on either side of large islands.

5.8.3. OTEC (Ocean Thermal Energy Conversion Plant)

Ocean thermal makes use of temperature gradients in a thermal (Rankine) cycle process. Ocean Thermal Energy Conversion plants must be situated where there is a difference of near about 40°F occurs throughout the year.

Depths of the Ocean must be abundant fairly close to shore-based facilities for economic operation. Floating plant ships could offer more versatility. Only small-scale versions have been made.

The largest Ocean thermal energy conversion plant so far is near to Japan, and that plant has the potential to produce a 100 kilowatts of the electricity. Another small-scale ocean thermal energy conversion power plant is off the shore of Hawaii, which produces near about fifty kilowatts of the electricity.

If a successful ocean thermal energy conversion is constructed, it is planned to generate up to two megawatts of the electricity. Nevertheless, a full-scale ocean thermal energy conversion plant can cost hundreds or thousands of the millions of dollars, and it would be very challenging task to construct.

5.8.4. Tidal Energy

Tidal energy is one of the oldest forms of energy utilized by the humans. Certainly, tide mills, in use on the Spanish, French, and British coasts, date back to the time period of 787 A.D. (Anno Domini). Tide mills are made up of a storage pond, filled by the incoming (flood) tide with the help of a sluice and deflated during the outgoing (ebb) tide with the help of a water wheel. The tides help to turn or spin the waterwheels, which in return generates mechanical power to mill grain.

Tidal power is non-polluting, reliable, and foreseeable. Tidal barrages, underwater tidal turbines such as wind turbines but runs with the help of the sea, and a diversity of machines connecting undersea currents are under development. Dissimilar to the wind and waves, tidal currents are completely foreseeable.

5.8.5. Wave Energy

Wave energy is an uneven in nature and oscillating at low-frequency energy source that can be converted to a 60 Hz frequency and can then be added to the electric utility grid. The energy in waves comes from the movement of the ocean and the changing heights and speed of the swells. Kinetic energy, the energy of motion, in waves, is tremendous (Figure 5.21).

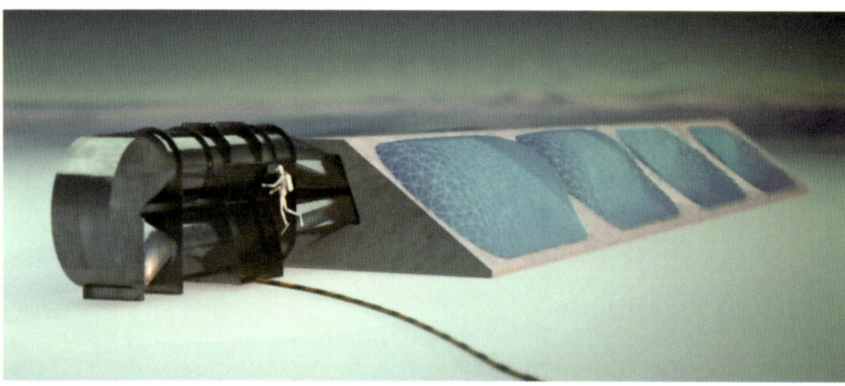

Figure 5.21: Wave energy.

Source: https://upload.wikimedia.org/wikipedia/commons/1/12/Bombora_mWave_Converter.jpg

An average 4 ft, 10 sec wave hitting a coast puts out more than 35,000 HP per mile of coast. Waves get their energy from the wind. Wind comes from solar energy. Waves collect, store, and convey this energy thousands of miles with little loss. As long as the sun shines, wave energy will never be exhausted. Wave energy can be fluctuating in terms of intensity, but it is available 24 hours a day, and 365 days a year.

REFERENCES

1. *Biomass at a Glance*, (n.d.). [eBook], p. 1. Available at: https://www.need.org/Files/curriculum/Energy%20At%20A%20Glance/BiomassAtAGlance_11x17.pdf (Accessed on 21 May 2019).
2. Davison, A., (n.d.). *Biomass Energy, Definition of Biomass Energy and the Types of Biomass Energy, Pros and Cons*. [online] Altenergy.org. Available at: http://www.altenergy.org/renewables/biomass.html (Accessed on 21 May 2019).
3. Eia.gov., (2018). *Geothermal Power Plants – Energy Explained, Your Guide to Understanding Energy – Energy Information Administration*. [online] Available at: https://www.eia.gov/energyexplained/index.php?page=geothermal_power_plants (Accessed on 21 May 2019).
4. *Geothermal Energy – Renewable Energy World*, (2019). Renewableenergyworld.com [online] Available at: https://www.renewableenergyworld.com/geothermal-energy/tech.html (Accessed on 21 May 2019).
5. *Geothermal Power & Electricity Production – Renewable Energy World*, (2019). Renewableenergyworld.com [online] Available at: https://www.renewableenergyworld.com/geothermal-energy/tech/geoelectricity.html (Accessed on 21 May 2019).
6. Hydropower.org., (2019). *Types of Hydropower | International Hydropower Association*. [online] Available at: https://www.hydropower.org/types-of-hydropower (Accessed on 21 May 2019).
7. Instalbiz.com., (2019). *Types of Ocean Energy – Ocean Energy Resource*. [online] Available at: http://www.instalbiz.com/resources/46-resource-types-of-ocean-energy_48.html (Accessed on 21 May 2019).
8. Islam, M., Hasanuzzaman, M., Rahim, N., Nahar, A., & Hosenuzzaman, M., (2014). Global renewable energy-based electricity generation and smart grid system for energy security. *The Scientific World Journal*, [online] pp.1–13. Available at: https://www.hindawi.com/journals/tswj/2014/197136/ (Accessed on 21 May 2019).
9. Marsh, J., (2018). *What is Solar Energy? Solar Power Explained | Energy Sage*. [online] Solar news. Available at: https://news.energysage.com/what-is-solar-energy/ (Accessed on 21 May 2019).
10. *Photovoltaic Systems-Renewable Energy World*, (2019). Renewableenergyworld.com [online] Available at: https://www.

renewableenergyworld.com/solar-energy/tech/solarpv.html (Accessed on 21 May 2019).

11. Planète Énergies, (2015). *The Different Forms of Ocean Energy*. [online] Available at: https://www.planete-energies.com/en/medias/close/different-forms-ocean-energy (Accessed on 21 May 2019).
12. Renewable solar energy, (2011). *What is Biomass | ReEnergy Holdings*. [online] Reenergyholdings.com Available at: https://www.reenergyholdings.com/renewable-energy/what-is-biomass/ (Accessed on 21 May 2019).
13. Renewableenergyworld.com, (2019). *Hydropower*. [online] Available at: https://www.renewableenergyworld.com/hydropower/tech.html (Accessed on 21 May 2019).
14. Renewableenergyworld.com, (2019). *Ocean Energy*. [online] Available at: https://www.renewableenergyworld.com/ocean-energy/tech.html (Accessed on 21 May 2019).
15. Renewableenergyworld.com, (2019). *Photovoltaic Systems*. [online] Available at: https://www.renewableenergyworld.com/solar-energy/tech/solarpv.html (Accessed on 21 May 2019).
16. Renewableenergyworld.com, (2019). *Renewable Energy Overview*. [online] Available at: https://www.renewableenergyworld.com/index/tech.html (Accessed on 21 May 2019).
17. *Solar Electricity – Renewable Energy World*, (2019). Renewableenergyworld.com [online] Available at: https://www.renewableenergyworld.com/solar-energy/tech/solarconcentrating.html (Accessed on 21 May 2019).
18. *Solar Hot Water – Renewable Energy World*, (2019). Renewableenergyworld.com [online] Available at: https://www.renewableenergyworld.com/solar-energy/tech/solarhotwater.html (Accessed on 21 May 2019).

6

Alternative Renewable Energy Sources and Its Challenges

CONTENTS

The chapter Alternative renewable energy sources and its challenges are important in order to understand the type of alternative renewable energy and its benefits. It also explains the difference between renewable energy and alternative renewable energy. It also discusses why there is a need arise to use alternative energy. It also explains that it is alternative renewable energy efficient enough in order to replace the existing fossil fuels and discusses the opinions of proponents as well as opponents. It further explains different challenges the individual and countries face at the time of using alternative renewable resources. It also discusses other challenges that discourage the use of alternative renewable resources. In the end, it discusses some reasons why the prices of the alternative energy resources are so high and what are the factors that contribute to an increase in prices.

6.1. INTRODUCTION

With the continuous increase in the prices of oil, concerns over preservation of energy for future and change in the climatic condition all have contributed to invest heavily in the development of alternatives to conventional oil. "Alternative energy" is generally categorized into two types:

- One includes substitute that can replace existing petroleum liquids as a source of energy (biodiesel, ethanol, dimethyl ether, biobutanol, tar sand, coal-to-liquids, and oil shale), both from fossil feedstocks and biomass; and
- The second one includes alternatives for creating and storing electric power (solar photovoltaic (PV), wind, tidal, solar thermal, fuel cells, biomass, and batteries).

There are many technological ways through which these alternative sources of energy can be developed, it varies, from distillation and gasification to bioreactors of algae and high-tech manufacturing of photon absorbing silicon panels. Some of the created energy are considered to be "green" or "clean," while on the other hand some of them such as coal-to-liquids and tar sands, have a more negative impact on the environment as compared to petroleum products from which they were replaced. Others, such as biofuels, are environment-friendly and have positive environmental impacts that offset potential carbon savings. Unlike the case of conventional fossil fuels, in which the landscape provide energy, and it takes about millions of years to form energy-dense solids through biomass, liquids, and gases are formed just through the means of transportation technology and extraction in a way to transport it from one places to another alternative source of energy generally relies upon modern infrastructure and engineered equipment for formation and conversion, because of this, it is mostly recognized as high-tech manufacturing process.

In addition, recent advancement in technology and new innovative techniques helps in making the formation of alternative source of energy easy; however, to deal with the technology in creation process is a complex task.

The entire process for creating alternative energy, starting from the collection of raw materials and goes until the manufacturing process, relies heavily on fossil fuel energy for performing various operations such as transportation, mining, and materials production. The primary challenge that the alternative energy faces is that how to supplant a fossil fuel–based supply chain with one driven by alternative energy forms themselves in order to break their reliance on a fossil fuel foundation.

The main area of discussion is how to minimize the financial cost of alternative energy as compared to traditional fossil fuels, often in comparison to their carbon footprints. This kind of reductionism to a simple monetary metric obscures the complex issues surrounding the potential viability, feasibility, scalability, and preference of choosing one technology over others. Although it is important to have enough financial resources to alternative energy, these are just a side support to mobilize different resources vital to produce energy.

At the level of physical requirements, analyzing the capability for developing alternative energy is a much more complex task since it is concerned with tackling the issue of end-use energy requirements, material scarcity and resource use trade-offs (including land and water). In the same way, it is often misrepresented that alternative energy can be used as an alternative to gas, oil, or coal—but this is not the perfect substitute for the same.

The use of alternative energy into our current energy system requires heavily investment to develop modern infrastructure and to purchase new equipment—along with the other necessary resources that are important in manufacturing process—mostly at a time when it is not easy to secure capital to make such investments. It thus results in raising concern of the stability of moving toward an alternative energy future with an assumption that the present available infrastructure and centralized energy system should be maintained.

Since the availability of alternative energy resources varies from location to location, it is important to examine different types of energy available at different locations. Each alternative has its own importance and uses; thus, it is important to identify reserves where the most used alternative energy is available in abundance quantity.

6.2. RELATIVELY NEW CONCEPTS FOR ALTERNATIVE RENEWABLE ENERGY RESOURCES

6.2.1. Biogas Digestion

Biogas digestion harnesses the methane gas that is produced during the time when organic waste breaks down in an anaerobic environment. Biogas digestion is a process in which biogas is extracted from sewage systems or from landfill sites. It is used to generate electricity, hot water, or heat. The methane gas that is extracted and then refined is an alternative source of energy and can be used for various products (Figure 6.1).

Figure 6.1: Biogas plant and container.

Source:https://cdn.pixabay.com/photo/2017/11/05/00/08/biogas-2919235_960_720.jpg

Reasons for interest in biogas digester system:

- This technology system to produce gas is more reliable than others.
- It is the best way to manage manure, resulting in tackling the issue related to odors and other environmental concerns.
- Government provides tax incentive to producers of biogas.
- It encourages and promotes clean energy.

6.2.2. Biomass Briquette

Biomass briquettes are a biofuel substitute to charcoal and coal. Biomass briquettes are generally formed by using the waste from forest and agricultural sector. The biomass formed from agricultural and forestry waste are originally low in density, and then it is converted into high-density biomass briquettes by using advance briquetting machine that uses binder less technique, and in this process, no chemical is involved, which means it is 100% natural (Figure 6.2).

Figure 6.2: Biomass briquettes.

Source: https://upload.wikimedia.org/wikipedia/commons/6/69/Straw-hay-briquettes.jpg

The primary raw materials that are used in creating biomass briquette are Sawdust, Mustard Stalks, Sugarcane Bagasse, Caster Seed Shells/Stalk, Cotton Stalks, Bamboo Dust, Wood Chips, Tea Waste, Tobacco Waste, Bajra cobs, Maize stalks, Paddy Straw, Arhar stalks, Sunflower Stalk, Wheat Straw, Soybean Husk, Coffee Husk, Rice Husk, Jute Sticks, Coir Pitch, Groundnut Shell Palm Husk, Barks & Straws, Veneer Residues, Pine

Niddle, Leaf, Seeds Cases, etc. Biomass Briquette are mainly used in all type of thermal application such as steam generation in boilers, in foundries and furnace (It can be used to melt metal at a time when melting point is less than 1000 d/Celsius), for heating purpose (it is used during the time of winter in commercial and residential sector to heat water or for cooking) drying process and in gasification plant replacing conventional solid fuels such as firewood and coal and liquid fuels such as Kerosene, Diesel, and Furnace Oil (FO), etc.

6.2.3. Biological Hydrogen Production

Hydrogen gas is an entirely clean-burning fuel; water is only byproduct of it. It also comprises of high density of energy as compared to other fuels because of its chemical structure.

$$2H_2 + O_2 \rightarrow 2H_2O + \text{High Energy}$$
$$\text{High Energy} + 2H_2O \rightarrow 2H_2 + O_2$$

This requires a high-energy input, resulting in making hydrogen less viable to produce commercially. The criteria to use biological vector in a way to split water, and thus produce hydrogen gas, would allow for the only energy input being solar radiation. Biological vectors mainly comprise of algae and bacteria. The entire process is commonly known as biological hydrogen production. It makes use of single-celled organisms in order to create hydrogen gas through the process of fermentation.

The environment in which no oxygen is available is called anaerobic environment, and in this type of environment, it is not possible for regular cellular respiration to take place and a process known as fermentation takes over. Hydrogen gas is a major by-product of this process. If it is possible to carry this process on a large scale, then nutrients, sunlight, and water could create hydrogen gas to be used as a dense source of energy.

6.2.4. Hydroelectricity

Hydroelectricity (also known as hydro) is a well-developed alternative renewable energy technology that is present in the world since past few decades (Figure 6.3).

Figure 6.3: Hydroelectricity power station.

Source: https://upload.wikimedia.org/wikipedia/en/4/46/Castaic_-_Front_of_Plant.JPG

Hydro makes uses of flowing water in a way to rotate a turbine that is attached to a generator in order to generate electricity. The extent to which electricity is produced relies on the speed and volume of water that flows above the turbine.

Hydroelectric power stations that run on a large-scale need dams to store the water in a required quantity to generate electricity. These dams are also useful in storing water for irrigation process and for drinking purpose. Maximum value attained from these dams by installing power stations, so that it can be used for multiple purposes.

It is important to note that hydroelectricity does not consume any water, as water flow to the river directly after generating electricity. While hydro plants can have very large capacities, the volume of electricity generated by it varies from time to time as it relies on the demand of electricity and on the monsoon season. Hydro can provide both baseload and peak load electricity, and hydro generators can start up and supply maximum power within 90 seconds.

6.2.5. Algae Fuel

It is a type of biofuel which is derived from algae. During photosynthesis process, organisms such as algae and others involved in photosynthesis collect direct sunlight and carbon dioxide and then convert it into biomass and oxygen.

The fuel is derived by putting the algae between two panes of glass. There are primarily three types of fuels that are created in this process. It includes biofuel (the natural "oil" derived from the algae), heat (produced during its growth cycle), and biomass (from the algae itself, as it is harvested upon maturity).

6.3. THE DIFFERENCE BETWEEN ALTERNATIVE AND RENEWABLE ENERGY

Although the primary objective of both renewable energy and alternative energy is to minimize the carbon emission level, there is a considerable difference between both of them. The basic difference between the two is that alternative energy is not unlimited in supply just like in the case of renewable energy, which, as the name suggests, is always available, similar to solar energy.

Now, let's examine the key differences between these two commonly mistaken and misinterpreted terms:

First of all, renewable energy is a natural occurring process as it does not require any person interference to produce these resources. Some of the examples of renewable energy include solar energy, wind energy, biomass resources, geothermal, and hydro resources.

One of the resources that are available in abundance quantity is solar energy. There is no point to discuss about the reason of abundance of solar energy as everyone knows about it. The sun is always shining at some point everywhere on Earth, making it the most abundant to harness. In addition, solar power also does not have any negative impact on earth's atmosphere, its environment, and wildlife.

While on the other hand, alternative energy does not comprises of resources such as solar power, but generally includes resources such as natural gas (that is mainly created by fracking or the injection of pressure in subterranean rock crevices), fuel cells, natural gas cogeneration, or any other waste energy that does not independently restore or reload but also contributed very less in carbon emission level. Oil cannot be considered as an alternative energy because it is the significant contributor to carbon emission level and also does not replenish naturally. Oil reserves present on the earth will diminish completely on one day, but it may few centuries. Thus, for the survival of our economy, ecosystem, and atmosphere, it is imperative to choose solutions like solar panel power.

6.4. NEED FOR ALTERNATIVE RENEWABLE ENERGY SOURCES

6.4.1. Energy Crisis

The energy crisis stems from the foreseeable end of the cycle of oil, gas, and coal, which, in addition, have been producing a considerable increase in greenhouse gases (GHG) (Figure 6.4).

Figure 6.4: Sustainable biofuel.

Source: https://upload.wikimedia.org/wikipedia/en/3/31/Pongamia_Pinnata_Seeds.jpg

In current few years, there have been many concerns raised by environment protection agencies about continuous change in climatic conditions, mainly due to the burning of coal and oil resources to produce energy in order to meet the huge demand of present generations. With the increase in the demand of these resources, the pressure on the fossil fuel increased, resulting in degrading the environment at a faster rate.

6.4.2. Causes for Energy Crisis

The global demand for energy consumption has been continuously increasing, and after few decades, the shortage of fossil fuels will quite observable. Therefore, it is important to have necessary reserves for future demand.

Present consumption pattern of today's generation primarily relies on the usage of non-renewable energy sources such as gas, oil, uranium, and coal. If the present rate of consumption continues, the first fossil fuel that will be soon finished is oil. According to survey recently conducted on oil reserves countries, it was found that there would be between 40 and 60 years of proven reserves of conventional oil. The estimated period to extract natural gas could be not more than 70 years, and for coal, the estimated reserves are for over a period of two centuries.

6.4.2.1. Overpopulation

No one can deny the fact that present demand on these nonrenewable resources is already high and with the increase in the population over a period of time – the world's population should reach nearly 10 billion people in 2050 – and the economic boom of growing areas, the pressure on these resources has been continuously increasing. The demand for the energy generally varies from one country to another. According to the International Energy Agency (IEA), the overall global demand for energy could increase by more than 50% by 2030 in the absence of public policies in this area.

6.4.2.2. Aging Infrastructure

It is generally noticed that the poor infrastructure of energy generating equipment results in scarcity of resources and energy shortage. Due to low budget and lack of knowledge about available advanced technology, most of energy producing companies keeps on using outdated equipment that limits energy production.

It is important to upgrade to modern infrastructure to ensure higher productivity and less wastage of resources, while on the other hand, implementation of high standard of equipment is essential for the security and safety of environment. There is a need to implement controlling measures in order to keep a check on energy producing companies.

6.4.2.3. Energy Waste

It is one of the most prevailing problems that is facing by everyone all over the world. Energy waste is the result of using energy resources inappropriately and without proper guidance and management. It can be defined as the wastage of resources such as fuels and electricity without predicting its impact on the resent as well as future generations.

It is important to take necessary actions on national as well as on an international level to minimize the wastage of energy and to develop a plan for energy savings. It requires the cooperation and coordination between various agencies and government to implement controlling measures and develop suitable plans to ensure sustainability of environment.

6.4.3. Effects of Energy Crisis

6.4.3.1. Environmental

The increase in the usage of traditional energy that is not environmentally friendly such as fossil fuels results in producing greenhouse gas emissions such as carbon dioxide (CO_2), which cause global warming and harming the environment and biodiversity. Therefore, it is not wrong to say that the energy crisis is closely linked to the environmental crisis.

6.4.3.2. Economic and Socio-Political

It is generally seen that energy security is one of the important points of concern for every local or national government all over the world. One of the primary reasons for this concern is that energy security is an important from the point of the overall growth in the economy. It is important for an individual country growth and development.

The energy crisis could thus have a dramatic impact on the global economy as it hinders the overall functioning in the economy. Energy plays an important role in manufacturing as well as transportation of goods and services, and thus, when energy markets fail, the business sector is the one that is mostly affected by this failure. Therefore, energy shortages and resulting economic factors may create socio-political issues.

6.4.4. Energy Transition to Alternative Renewable Energy Sources

Unlike fossil fuels, there are many other energy sources that are totally renewable and do not contribute to the GHG. These are clean, environmentally friendly, and ensure sustainability in the environment. It includes solar energy, wind energy, hydropower, geothermal energy, and biomass energy.

6.5. BENEFITS OF ALTERNATIVE RENEWABLE ENERGY USE

Wind turbines and solar panels are now commonly used by every country because of the benefits it provides in terms of healthy environment, improved health, sustainability, and important for the growth of the economy (Figure 6.5).

Various positive impact of clean energy, including the benefits of solar, wind, hydroelectric, geothermal, and biomass are explained below.

Figure 6.5: Installation of solar panel to generate electricity.

Source: http://navylive.dodlive.mil/files/2016/05/120719-N-RI884-240.jpg

6.5.1. Less Global Warming

Human activity has a negative impact on our earth as it results in producing carbon dioxide and other global warming emissions at a large scale. These gases act like a blanket, trapping heat. It may have significant and harmful impacts on environment, from drought, more frequent storms, to sea level rise and extinction. According to a recent survey conducted in the United States, it was observed that electricity sector contributed to approximately 29% of total global warming emission.

In contrast, most of the renewable energy sources produce very little to no global warming emissions at all. Even when we determine the total emission from clean energy during its production process, it is still very small in amount.

The comparison becomes even clearer when it is analyzed with its numbers. The process of burning natural gas to produce electricity produce between 0.6 and 2 pounds of carbon dioxide equivalent per kilowatt-hour (CO_2E/kWh); coal emits between 1.4 and 3.6 pounds of CO_2E/kWh. While on the other hand, wind contributes to only 0.02 to 0.04 pounds of CO_2E/kWh on a life-cycle basis; geothermal from 0.1 to 0.2, and solar from 0.07 to 0.2.

6.5.2. Improved Public Health

The air and water pollution released by coal and natural gas industries results in causing neurological damage, breathing problems, cancer, heart attacks, premature death, and a number of other dangerous problems. Pollution is the major problem of every country. In a recent study by Harvard University, it was estimated that the life cycle costs and public health effects of coal to be an approximately 74.6 billion every year (Figure 6.6).

Most of these problems exist because of the burning of fossil fuels, and it is not the case with clean energy technologies. Solar, wind, and hydroelectric systems generate electricity without any release of air pollution emissions. Biomass and geothermal produce very little air pollutants but these are far less as compared to emissions generated by coal- and natural gas-fired power plants.

Additionally, the process that generate solar and wind energy does not require the use of water for production, thus do not result in effecting water resources and do not release any remains that come in contact with water, thus avoiding water pollution, while on the other hand, during the production process of coal mining and natural gas, various harmful remains come in contact with water and air, thus result in affecting the environment.

Figure 6.6: Producing electricity to fulfill energy needs.

Source: https://upload.wikimedia.org/wikipedia/commons/3/32/Wind_Power_in_Rural_Iowa.jpg

6.5.3. Jobs and Other Economic Benefits

Compared with fossil fuel technologies, that are capital intensive and automotive, renewable energy industry are labor intensive, which means more labor force are required to carry out its operations. For example, solar panels need labor force to install them on rooftop, and wind farms need technicians for installation and maintenance. Thus, it can be said that a greater number of jobs will be created in each number of electricity produced from renewable resources as compared to fossil fuels.

According to a survey recently conducted in united states, it was found that there are approximately 10,000 people that are employed in the wind industry, including project development, manufacturing, construction, and turbine installation, transportation, and logistics, operations, and maintenance, and legal, financial, and consulting services.

Other renewable energy technologies such as solar power even employ more than 2,60,000 people, including jobs in production process, sales, and installation of solar panels. The hydroelectric power industry employed approximately 66,000 people in 2017; the geothermal industry employed 5,800 people.

6.6. CAN ALTERNATIVE ENERGY EFFECTIVELY REPLACE FOSSIL FUELS?

6.6.1. Proponents Point of View

- One of the proponents is of the view that it will take few decades to completely leave behind fossil fuels. But it is not impossible, if proper research and innovation done in this field, and then it is not difficult to achieve the objective.

But it required huge cost in the process of transition, not only in the terms of regulations and money, but also in the terms of behavior of people and their usage pattern. It will take some time, but it provides ample benefits and change the way we live. It is commonly believed that the best way to start the transitions is to replace the coal with solar and wind power for electricity generation. According to certain research, it was found that the true all renewable economy will be completely different than what we see today. It will probably be a conserver economy rather than a consumer economy. It will also likely feature far less economic inequality.

- Another proponent is of the view that the day will soon come, when renewable energy such as solar, wind, geothermal, and others that are still in development replace fossil fuel and used as a primary source of world energy.

The process of transition completely relies upon improvement in the technology and in pricing to acquire such technology. As the efficiency of wind and solar system is increasing, along with the reduction in cost by producing it at a large scale and thus giving the benefits directly to the customers, resulting in fall in prices of renewable energy globally.

However, it is wrong to believe that renewable energy replaces fossil fuel in next few years. Nevertheless, renewables are likely to replace fossil fuels as the dominant source of electrical power well before mid-century as well as make giant strides in other areas such as transportation."

- However it is difficult and expensive to shift completely too green energy, but it won't take long time when initial investment in transition will start to set off by lowering the electricity bills as well as gasoline bills. It will not take much effort as well as cost to do cost more to install solar arrays and build wind farms. But once they become operational, there will be less maintenance cost as compared to conventional power on a yearly basis. No more regular fuel bills and only minimal expenses for upkeep of solar, as the wear and tear concept are not applicable in case of solar power because all parts are immovable and there is no need to replace them.

The price argument is also fallacious because it is not possible to determine the actual cost of power that we used as it includes the cost to the human health and to the environment of the carbon pollution that fossil-fuel mining and burning generates. There is need to promote and encourage those technologies that help us to put the brake on destructive climate change, rather than feeding the unsustainable fossil-fuel habit that is driving it.

- According to Alfred W. Crosby, PhD, Professor Emeritus of History, Geography, and American Studies at the University of Texas at Austin, "Alternative sources of energy can become a satisfactory substitute for fossil fuels if we put as much effort and genius in the effort as we did in producing the first atomic bomb.

The most satisfactory single alternative would be hydrogen fusion, but that quasi-miracle may be beyond our capability. We may discover that wind, solar, biomass, etc., all piled on top of each other, may have to do, but their success may turn out to require an effort that started a generation ago. Essential to any and all success is the realization on our part that we may be able to do anything, which includes fail."

- Renewable energy, can, and indeed must replace both fossil fuel and nuclear power as quickly as possible if the world wants to avoid further degradation of environment and adverse climatic conditions.

According to the 2nd edition of the Energy Revolution contains this statement: "The amount of energy that can be accessed with current technologies supplies a total of 5.9 times the global demand for energy." The remaining report believed that it is not possible to clear the path to a clean energy future without tackling the current political and economic constraints.

According to an expert, there is need to encourage politicians to actively participate in the process of transition and enhance their knowledge, energy, and courage if they want paradigm shift from fossil fuel and nuclear eras.

Is it possible to make that leap with available technology? Yes, geothermal energy, Tidal power, cogeneration, and biomass combined with conservation are some of the resources yet to be explored. Presently, in some countries, all electricity is generated by a combination of renewable carbon-free and nuclear-free technologies.

6.6.2. Opponents Point of View

It is important to note that gasoline provides approximately 95% of the fuel used in the transportation sector. Every transport mode – whether it is trucks, cars, buses, trains, or aircraft and marine vessels, and – relies completely on petroleum fuels. Only natural gas liquids and, recently because of some restriction, ethanol have made small inroads in the dominant share held by oil.

In addition, according to some surveys and studies, it was found that this trend will continue even till the year 2035 to 2040. Proponents of the all-renewable future proved themselves wrong in their own assumption. According to them, in 2018 there would be fall in the prices of gasoline as people shift to cleaner energy, but that is not the case, still now the price of the oil is more than $70 per barrel, and natural gas and coal prices are surging.

In fact, the gasoline becomes more competitive, and the demand for gasoline is continuously increasing. In such a world, it can be said that the importance of renewables will grow but gradually. Also, it is found that due to increase in the production of oil, the international oil price will fall to $40 per barrel and the dramatic slumps in natural gas and coal prices in some countries makes the alternative energy less competitive and it result in demotivate people to shift their modes of transport to renewable energy as it includes huge investment in transition.

So, is the solar revolution finally here? Not quite. Even after a decade of rampant growth solar energy still not used at a large scale. In fact, solar merely equals the amount of electricity produced all over the world by burning natural gas captured from landfills.

The large number of resources used today is still traditional. Oil still reigns supreme at 36 quadrillion Btu, nuclear 8, natural gas at 26 quads, biomass, and hydropower bring up the rear at 2.7 and 2.6 quads. Wind is just 1.5 quads. And coal, largest contributor to air pollution level stood at 19 quads. That's approximately eight times of all the solar and wind energy produced by the entire world.

For all the talk of 'grid parity,' the bitter truth is that it will take few decades to renewable energy to produce at a large scale and satisfy the demand of entire world. While on the other hand, the continuous fall in the prices of coal and other fossil fuels will throw another challenge to renewable energy to be accepted at wider scale.

In addition, due to continuous advancement in the technology that extracts coal, it become immensely cleaner over the past generation. And also new and efficient technique will be found to extract energy from coal without producing hazardous products that result in degradation of environment. Its mobility and capacity to generate more energy for less storage used make it more efficient than renewable energy. In short, it can be said that due to its reliability and usability, coal will remain accepted globally.

According to Clive Best, PhD, a former physicist, "Modern society depends on always available power. If power goes down, then society stops. There are no phones, no internet, no ATMs, no refrigeration, no sewage pumps – nothing, and if a large city like London is without power for more than 12 hours rioting and looting would quickly take hold. It is, therefore, inconceivable not to ensure that we always have reliable energy.

So, an energy plan for the UK must be able to meet demand even on the coldest evening of the year in winter with no wind and no solar. For this reason, Renewable energy can never under any realistic scenario meet that target. To imagine that battery prices could fall enough to make wind and solar backup, such enormous power demands is simply a delusion."

6.7. CHALLENGES IN ALTERNATIVE ENERGY SOURCES

6.7.1. Scalability and Timing

In order to ensure that the primary purpose of an alternative energy source to be achieved, it is important to supply the necessary energy in a particular time frame in the required quantity and at a reasonable cost. Many alternatives have been successfully demonstrated at the small scale (cellulosic ethanol, algae-based diesel, thin-film solar, biobutanol), but demonstration scale does not self-sufficient in providing an indication of the potential for large-scale production.

In the same way, as the production of alternative energy relies heavily on manufacturing, engineering, and development of modern equipment and advance manufacturing processes for its production. The total output generated from this entire process, produced step by step as it as in each process or function, some value adds to the product and it generally relies on timely procurement of the input energy, necessary resources, and other required input materials.

This difference between "extraction" of fossil fuels and "production" of alternative energy can result in marked constraints on the ability to increase the production of an alternative energy source, as it is needed.

6.7.2. Commercialization

There are several issues in the commercialization of the alternative renewable resource. Whether it is scalability issues, issues related to obtaining license for commercialization, the time it takes for fully commercialize the entire process. As the technique used in the production of alternative energy is complex, it is not easy for any company to simply start the production of alternative energy.

It is worth noticing that the average time frame between a laboratory demonstration of feasibility and large-scale commercialization is from 20 to 25 years. The entire processes involve in the production of alternative energy need to be optimized, monitored, demonstration tests performed, patents developed, environmental impacts assessed, pilot plants built and evaluated, comply with the industry standards, and design, engineering, sitting, financing, economic, and other studies were undertaken.

6.7.3. Substitutability

It is generally observed that an alternative source of energy directly comes into the economy and initially considered as the best substitute for the prevailing energy in the market. Some of the introduced alternative energy does not even require any change in the current infrastructure sector. For example, in the case of introduction of electricity vehicles, which are the best alternative for already available gasoline, run vehicles, it comes along with other major challenges in the adaption of electric vehicle on a large scale.

Although one cannot deny the fact that solar power that is required to charge electric vehicle is available in abundance quantity, but the process that start from manufacturing of electric vehicle to successful run the vehicle on roads is quite complicated and challenging.

Manufacturing of electric vehicle at a large scale require huge investment in research and development, modernization of infrastructure including retooling factories to produce the vehicles, developing a large-scale battery industry and installation of electric vehicle charging station, extraction of lithium-ion resources (which is important to produce electric batteries and there are only few reserves of lithium ion), building a maintenance and spare parts industry, integrating “smart grid” monitoring and control software and equipment, and of course, constructing additional generation and transmission capacity. All of these resources or activity to run the electric vehicle project is quite costly.

While on the other hand, wind, and solar power electricity project demands additional infrastructure and other necessary resources to run the project successfully; it is important to choose a place where enough resources to produce wind and solar electricity are available, even maybe it is far away from populous areas. Thus, huge investment required to develop transportation infrastructure sector in order to mobilize the resources from one place to another, where large number of populations lives.

Currently, it is possible to integrate ethanol with gasoline and used directly in the vehicle as an alternative to gasoline, but there are few challenges to run it effectively as it generally absorbs water and there are high oxygen content present in it. It requires making certain changes in existing pipeline system to use ethanol in place of fuels, and it will take certain time to implement this technique successfully. While alternative energy forms may provide the same energy services as another form, they rarely substitute directly, and these additional material costs need to be considered.

6.7.4. Material Input Requirements

The key input required in the production of alternative energy is not money, but resources and energy; the quality and quantity of the energy and resources may, in turn, impact the overall production volume and the cost involved in the entire process as well as the feasibility of an alternative. It is quite observable in energy projects that relies heavily on modern technologies and techniques in manufacturing process and rarely deals with earth elements.

For example, fuel cells require palladium, platinum, and rare earth elements. Solar PV technology requires gallium, and in some forms, indium. Electric batteries are formed by using lithium resources. In addition, the present technology that is designed to save energy, such as LED or organic LED (OLED) lighting, requires the rare earths elements such as gallium and indium. Determining the value of alternative energy only in terms of monetary terms results in impeding the resource and energy inputs required.

Implementation of necessary energy technologies and some advance or modern technologies results in increasing the demand of certain metals significantly, even more than its production level.

The production of alternative energy not only depends on different types and volume of resource inputs, but also on fossil fuels that are important for the mining of raw materials, manufacturing, transport, maintenance, construction, and decommissioning. Presently, no alternative energy exists without fossil fuel inputs, and it is not possible to produce alternative energy without the use of fossil fuel as the equipment used in the production of alternative energy, first of all, developed by the use of fossil fuel. In this regard, alternative energy serves as a supplement to the fossil fuel base, and its input requirements may constrain its development in cases of either material or energy scarcity.

6.7.5. Intermittency

It is general belief among the modern societies that that gas will supply only when a knob is turned on, electrons will flow when a switch is flipped, and liquid fuel will supply when the pump handle is turned on. The continuous supply of all these resources is possible only because of our consistent practice of exploitation of large stores of fossil fuels, which are the result of millions of years of intermittent sunlight concentrated into a continuously extractable source of energy.

Production of alternative energies such as wind or solar power only possible when the Sun shines or the wind blows, regularly or after a certain period and even the fuels based upon biomass depend on season of harvests of crops. Employing these energy forms into our present available system comes along with various challenges of complying with the demand, and it is always doubtful that these irregular supply of energy resources meet the demand of future generation as presently the pressure on these resources are quite high and demand is more than the supply.

The best way to deal with the intermittency is to use the storage options; that is, involving in research and development to develop new technologies that can effectively store wind or solar power generated over a period of time. Many potential approaches have been planned and tested, including batteries, compressed air storage, and the use of molten salts in solar thermal plants. The downside of this approach is that lot of energy wasted during storage and transfer of energy, and also these technologies have limited capacity to store energy.

6.7.6. Energy Density

Energy density refers to the total amount of energy that can be stored in a single unit at a time. In other terms, it can be defined as total amount of energy per unit of mass (weight), or the amount of energy per unit of volume. Energy density plays a major role in usability of fuels. In the seventeenth and eighteenth centuries, the use of coal is most preferable as it provide twice the energy as provided by wood for the same weight of material.

In the same way, in the past few decades, there was considerable shift from coal to petroleum-powered ships as the petroleum provided twice the energy for the same space, allowing ships to go farther without having to stop for refueling. Even when used in a motor vehicle's inefficient internal combustion engine, a kilogram of highly energy-dense gasoline—about six cups—allows us to move 3,000 pounds of metal roughly 11 miles.

The major downside of low energy density is that for the same productivity or efficiency, large space required storing energy. There are many different types of storage technologies, and alternative energies that are mainly categorized into low energy densities, and their deployment will result in higher levels of resource consumption.

Lithium-ion batteries that are vital for the success of electric vehicles can contain only 0.5 megajoules per kilogram (MJ/kg) of battery as compared to 46 MJ/kg in case of gasoline. There has been rapid growth in the advancement of battery technology, but they all come up against the theoretical limit of energy density in batteries of only 3 MJ/kg.

6.7.7. Energy Return on Investment

The complexity of our economy and society is a function of the amount of net energy we have available. "Net energy" can be simply defined as the total amount of energy left after satisfying the demand all over the world. It is impossible to save the energy that is used during the production of energy, but the energy that is used apart from producing energy can be saved to some extent such as those used in transport, industrial, commercial, residential, agricultural, and military activities.

The ratio of the amount of energy used into production of energy and the amount of energy we produce is called energy returns on investment (EROI). EROI can be very high (e.g., 100:1, or 100 units of energy produced for every one unit used to produce it—an "energy source"), or low (0.8:1, or only 0.8 units of energy produced for every one unit used in production—an "energy sink"). For the success and growth of an economy, it is important to have energy sources rather than energy sinks and the percentage of EROI for an energy source is a key indicator in a way to check its overall productivity and its contribution in the economy. The total net energy availability varies from time to time and from one location to another.

In the present modern society that totally relies upon solar power (sun, biomass, water power, and the animals that depended on biomass), while on the other hand, in the seventeenth and early eighteenth centuries, the total available net energy was quite low and primarily relies upon the food surpluses provided by farmers and also at that time, approximately 85 to 90% of the population engage in the activity of energy production.

As extraction of oil, coal, and natural gas was started at a rapid rate during the nineteenth and twentieth century's, society was increasingly able to substitute the energy from fossil fuels for manual or animal labor, thereby allowing an even larger proportion of society to free from directly engage in the activity energy production.

In 1870, it was observed that approximately 70% of the U.S. population was involved either directly or indirectly in farming activity; in the present era, this figure has been fallen significantly to only just 2% or even less, and mostly all the activities involved in the agricultural production now completely relies on petroleum or natural gas. This type of trend has also been seen in other energy sectors: presently, there are only less than 0.5% of the U.S. population (approximately 710,000 people) that is directly involved in the process of oil and gas extraction, coal mining, pipeline transport, petroleum refining, and power generation, transmission, and distribution.

The challenge of a transition to alternative energy, then, is whether it is possible to store this surplus energy, and thus whether the type of social and economic specialization that is presently available can be maintained.

According to one study conducted recently, it was estimated that the minimum EROI for the maintenance of industrial society is 5:1, which means that only 20% economic and social resources can be contributed to the production of energy without altering the structure of industrial society. It is generally seen that most of the energy sources have low EROI values. A high EROI is not enough to certify that the present structure of modern society and economies can be maintained, but it is a prerequisite.

6.8. OTHER CHALLENGES EXIST IN THE REAL MARKET CONDITIONS

6.8.1. The End to Government Incentives

The tax incentive schemes that are provided to the producers of alternative source of energy are continuously diminishing. In some countries such as in U.S.A., the scheme expired in the year 2013. This is one of the major challenges that is facing by the producers of alternative energy as huge investment required in the research and development of alternative energy system. Furthermore, due to advancement in technology, they required new technology to extract the resources at a rapid rate to comply with the present demand, and these technologies require huge financial resources to acquire.

In addition, lack of incentive scheme further hinder innovation in this sector as there is no motivation to do so. It also results in delay of present project and fear of massive job cuts. In most of the developing countries, where advanced technology and enough resources are not available to extract the alternative energy resources, it is important to provide proper incentive scheme to these countries. The policymakers must also analyze that is benefits generated from these schemes to the economy outweigh the subsidies given in the schemes.

According to some experts such as Mike Ryan, co-founder of Pantera Energy, alternative energy solutions provider specializing in geothermal systems, asserts that projects across the industry will be hurt by the looming deadline. "Industries like solar and wind will be difficult to wean off of the government incentives. Unless the technology can get to a point of affordability without the tax credits, many of the alternative energies will severely suffer."

6.8.2. Falling Natural Gas Prices Lessen Demand

As the demand of natural gas has been continuously increasing and at the same time with the increase in the supply, the cost of the natural gas is at its lowest level. Analysts having expertise in this field are of the view that the price of natural gas will continue to fall in the future also.

According to a study conducted by Massachusetts Institute of Technology, it was found that with increase in the supply of natural gas and decrease in its prices, the demand for renewable energy has been continuously decreasing. Most of the consumers are not changing their vehicle or other machinery that run on natural gas with Machinery that run on renewable resources as the first one become economical to run and the second one requires huge investment to shift from one means to other.

The booming shale gas industry is being partially blamed for several alternative energy industries cutting planned projects. Many alternative energy producing industries cannot contribute with the low cost of natural gas industries as they are well established and have enough resources to implement new techniques to further cut the prices of natural gas. And, with the end of the renewable energy tax incentives nearing, the pendulum continues to swing in the favor of natural gas.

6.8.3. Awareness Is Key

It is important for alternative energy companies to continuously raise awareness among people about natural gas and other fossil fuels and the impact the industry will have on the Earth's carbon footprint. The government, environment agencies, businesses, and consumers must ensure that the benefits generated from the alternative renewable energy must outweigh the cost of producing these resources.

The government should also actively participate to promote the use of alternative energy resources and to provide the tax incentives to the producers and consumers of alternative energy in order to defray the upfront costs of implementing such technologies. "With incentives, the return on investment is much faster."

Despite the number of challenges, the renewable energy industry is facing; currently, many people are still of the view that it is the future for the energy market. With continued "green" pressure, the country is making strides to be more eco-friendly, and alternative energy resources will remain a viable option.

6.9. WHY IS ALTERNATIVE RENEWABLE ENERGY SO EXPENSIVE?

It is mostly believed that carbon emissions from power stations are a significant contributor to change in the climatic conditions. These days a fiercer argument is over what to do about it. Many governments are investing heavily into renewable sources of electricity, such as solar farms, wind turbines, geothermal plants, hydroelectric. But countries that produce renewable energy on a large scale, such as Germany and Denmark, also have the highest energy prices all over the world. In Britain, the cost of electricity generated from wind farms is almost twice as compared to that from traditional sources; while on the other hand, solar power is dearer. What makes it so costly?

Enthusiasts have been using wind turbines to produce electricity since the year 1880s, but the plan to construct wind farms at a large scale came into effect only in the late 1970s. Utility-scale solar and other renewable generation is more recent still. Despite the subsidies and incentives provided by the government, there are only few industries involved in the production of alternative energy resources.

For example, in Britain, almost all wind turbines are produced by a single firm. Supply-chain bottlenecks have frustrated governments scrabbling to install new renewable capacity. And as compared with traditional power stations, the present available renewable generators are economical to run but are too expensive to construct, resulting in making them more vulnerable to changes in the cost of capital.

A more fundamental challenge is that renewable generators also impose costs on the wider electricity grid. The suitable sites are also far away from the big cities (on French lakes, Scottish hillsides, or American deserts), which makes them expensive to connect. It is also observed that most of the renewable generators produce power only after certain intervals, which means power generated only when sun shines or when wind blows.

Thus, it means that those countries that completely rely upon renewable energy for their energy need also required maintaining traditional kind of power generation in a way to ready for the peak time or during monsoon season when sunshine is almost unavailable. And energy from these stations also becomes more expensive because they may not run at full-blast. In addition, the government needs to maintain both type of stations which addition the cost significantly.

The high cost of renewable generators obstructs efforts to tackle climate change, even when governments dig deep to fund them. One of the major drawbacks of the sharp increase in the energy prices is that it enables manufacturers to set up in less "green" countries, which might mean citizens end up consuming more carbon, through imports.

Another important concern is that governments may continuously allow the functioning of dirty coal plants in order to serve as a back-up when productivity from renewable resources is low—or when over-ambitious renewable roll-outs run out of steam. But for now, the primary danger of high renewable costs results in selecting controversial alternatives.

The price of nuclear power has been increasing continuously since the past few decades, but still, its prices are far less as compared to renewable generation. The contribution to pollution level from Gas-fired power stations is approximately half than coal-fired ones. Building more of them could provide a cheaper way for countries to cut emissions in the short term and buy renewable operators time to bring their costs under control.

6.10. CONCLUSION

In the end, it is concluded that it is not possible to shift from the fossil fuels to alternative renewable sources entirely. Furthermore, the continuous fall in prices of gasoline discourage to use clean energy technology in vehicles. It is important for government of various countries and agencies to work collectively in order to faster implement renewable energy technology.

It means achieving cooperation and transparency between various departments to accomplish common goal. It requires further innovation and advancement in the technology to shift to alternative renewable sources. It will take some time to implement clean energy technology at a large scale.

REFERENCES

1. 3esavers.com, (2019). *What is Biomass Briquette*? [online] Available at: http://www.3esavers.com/about-biomass/what-is-biomass-briquette.html (Accessed on 21 May 2019).
2. Alternativeenergy.procon.org (2019). *Can Alternative Energy Effectively Replace Fossil Fuels? – Alternative Energy – ProCon.org*. [online] Available at: https://alternativeenergy.procon.org/view.answers.php?questionID=001244 (Accessed on 21 May 2019).
3. Birmingham.ac.uk., (2019). *Biological Hydrogen Production – University of Birmingham*. [online] Available at: https://www.birmingham.ac.uk/research/activity/bio-hydrogen/index.aspx (Accessed on 21 May 2019).
4. Business trends and insights, (2019). *Challenges to Energy: A Push for Renewable Resources*. [online] Available at: https://www.americanexpress.com/en-us/business/trends-and-insights/articles/challenges-to-energy-a-push-for-renewable-resources/ (Accessed on 21 May 2019).
5. Cleanenergycouncil.org.au., (2019). *Hydro | Clean Energy Council*. [online] Available at: https://www.cleanenergycouncil.org.au/resources/technologies/hydroelectricity (Accessed on 21 May 2019).
6. Davison, A., (2019). *Alternative Energy – Wind, Solar, Hydro and Other Alt Energy Sources for Commercial and Home Power*. [online] Altenergy.org. Available at: http://www.altenergy.org/ (Accessed on 21 May 2019).
7. Energy-reality.org (2019). [online] Available at: http://energy-reality.org/wp-content/uploads/2013/03/10_Alternative-Energy-Challenges_R2_012913.pdf (Accessed on 21 May 2019).
8. Impulse, S., (2019). *Solutions to the Energy Crisis: How to Achieve Sustainable Energy?* [online] Solarimpulse.com. Available at: https://solarimpulse.com/energy-crisis-solutions (Accessed on 21 May 2019).
9. Resilience, (2019). *Nine Challenges of Alternative Energy – Resilience*. [online] Available at: https://www.resilience.org/stories/2010-08-12/nine-challenges-alternative-energy/ (Accessed on 21 May 2019).
10. Saveonenergy.com., (2019). [online] Available at: https://www.saveonenergy.com/learning-center/post/what-you-need-to-know-about-alternative-energy-sources/ (Accessed on 21 May 2019).

11. Thakur, (2019). *What is Biogas Digestion?* [online] Slideshare.net. Available at: https://www.slideshare.net/nareshthakur75436/what-is-biogas-digestion (Accessed on 21 May 2019).
12. The economist, (2019). *Why is Renewable Energy So Expensive?* [online] Available at: https://www.economist.com/the-economist-explains/2014/01/05/why-is-renewable-energy-so-expensive (Accessed on 21 May 2019).
13. Union of concerned scientists, (2019). *Benefits of Renewable Energy Use*. [online] Available at: https://www.ucsusa.org/clean-energy/renewable-energy/public-benefits-of-renewable-power (Accessed on 21 May 2019).

7

Non-Renewable Resources and the Issues

CONTENTS

Since the ages, the non-renewable resources are the main source of the energy for the whole world. Every action on Earth is being fueled by the fuel that is being derived from the non-renewable resources of the energy. The fossil fuels are the main source of the energy amongst all the other sources of energy. Coal, oil, and natural gas comprise the main portion of the Non-Renewable Resources of energy. These resources have been used to such an extent that there have been some issues for the future and sustainability. Also, the main concern of the pollution dwells with the usage of non-renewable resources.

7.1. INTRODUCTION

The non-renewable energy resources originate from the sources that will run out by the time. Or the resources that will not be replenished or refilled in the lifetimes—or even in many, many eras.

Majority of the non-renewable energy sources are fossil fuels such as coal, petroleum, and natural gas. Carbon is one of the primary elements in fossil fuels among all. Because of this reason, the time period in which the fossil fuels formed (about 360–300 million years ago) is known as the Carboniferous period. This time period was near about 360 to 300 million years ago (Figure 7.1).

All the other fossil fuels are formed in a similar way like that of the others. Earth had a dissimilar landscape. This was at the time when even the dinosaurs did not exist on the earth. The time period was about hundreds of millions of years ago. The earth landscape at that time was covered with the wide, shallow seas and swampy forests.

There exist several plants that grew in these early swamplands. These are Plants, algae, and plankton. These plants, algae, and plankton absorb the sunlight. With the help of the sunlight, they created energy through the process of photosynthesis. At the time of their death, the organisms were drifted to the bottommost of the sea or lake. There still exists the energy that is stored in the plants and animals when they died.

Figure 7.1: There are many types of non-renewable resources.

Sources: https://upload.wikimedia.org/wikipedia/commons/0/01/Nuclear_power_plant_Isar_2.jpg

By the time, the dead plants were crumpled underneath the seabed. There are other elements such as the rocks, and the other sediment are heaped on the upper part of them. These elements are making high heat and pressure underground. In this kind of surrounding, the leftovers of the plant and animal are turned into fossil fuels such as coal, natural gas, and petroleum by the end of the time. In the present time, there are vast underground pockets of these non-renewable sources of energy all over the world. These underground pockets are known as the reservoirs.

On the other hand, the non-renewable resources are originated in the ground. There exists the fixed quantity of the resources. They are not the living things and also, at some of the time they are difficult to find. These nonrenewal resources do not regrow, and these are the resources that are not substituted or renewed. These resources comprise of the fossil fuels that afterburn for energy. These are the natural gas, coal, and oil. The minerals that are used for production of metals are also non-renewable natural resources. On the other hand, the nonrenewable natural resources are the substances that take a much longer time as compared to the natural life of the person in order to get replaced. In addition to it, these resources can take millions of years for the process of formation.

The type of the fossil fuels such as oil, coal, and gas will not last forever. These fossil fuels are nonrenewable in the nature. People are finding it difficult to search for the new fuels which are clean and will help in providing the power that is required. There are several numbers of renewable resources that provide a kind of hope for the upcoming times. These resources are wind, solar, and hydrogen power.

People all over the world make use of both the kind of natural resources. These people use these resources in order to produce the things as per their needs and requirement. The households, the clothing, plastics, and foods are all made from the natural resources.

The dissimilar things are originated from the different substances. For instance, the home is in a building. The Buildings on the other side are made from the substances like wood and minerals. Wood comes from the trees. On the other hand, the minerals are extracted from the ground. Bricks, cement, and metals are all made from minerals. How about the clothes that people are wearing? Almost all of the clothing is made from cotton, polyester, or nylon.

Cotton is made from the cotton plants. Polyester and nylon are made from oil. On the other hand, oil is also used in the formation of Plastics. How about your food that is consumed by the people all over the world? People in the world do the intake of grains, fruits, and other parts of plants. On the other hand, people may also enjoy the consumption of the dairy products and meat from animals. Everything that is existing, and that is used is made from a natural resource.

There is a discussion about the resources which are renewable, and which are non-renewable. Resources like coal, rocks, oil, minerals, and many more are the renewable resources or the nonrenewable resources.

The usage of all the existing natural resources should be wise. The people should focus on the conservation of the natural resources. The conservation of the natural resources means that people should not ruin or waste the things. Particularly, this is very true as well as important for the nonrenewable resources. Even though there are some renewable natural resources which can run out if they are all killed or if they are overused. So, it is very important for everyone all over the world to focus on protecting the natural resources from the pollution. Many activities can be the reason for the pollution. So, there is a need for proper measures in order to protect the environment and the resources from the pollution.

Pollution can rise when people put the dangerous chemicals and other things into nature that have a harmful effect on it. Oil is dropped into the water, toxic chemicals in the air, or the garbage is dumped on the side of the road, which gives rise to the pollution. These were the few examples of the problem.

There are nine most important areas of energy resources. These further falls into two categories, and one is nonrenewable and renewable. The non-renewable energy resources are like coal, nuclear, oil, and natural gas. These non-renewable energy resources are obtainable in the restricted supplies. Generally, this is because of the reason that these types of resources takes a longer time in order to replenish.

But on the other hand, the renewable resources are restocked naturally. And this happens over relatively short periods of time as compared to that of non-renewable resources. The five most important renewable energy resources are solar, wind, water (hydro), biomass, and geothermal.

7.2. CAN THE HUMAN RACE DWELL ONLY ON RENEWABLE ENERGY?

Since the very beginning of the human race, the people have used the renewable sources of energy in order to live. They make use of many resources for their survival. For instance, they make use of wood for cooking and heating, and they make use of wind and water for milling grain, and the use of solar for lighting fires.

Near about 150 years ago, the people made the technology to abstract energy from the fossilized leftovers of the plants and animals that were present in the olden times. These kinds of resources are super-rich, but on the same side, these resources are the limited sources of energy. These resources are coal, oil, and natural gas. These super-rich resources quickly replace the wood, wind, solar, and water as the most important sources of fuel (Figure 7.2).

Figure 7.2: It is important for the mankind to find good sources of energy.

Sources: https://upload.wikimedia.org/wikipedia/commons/1/13/Turbiny_wiatrowe_w_Szwecji%2C_2011_ubt.JPG

In the present energy marketplace, the fossil fuels make up the big share. Even though there are new hopeful renewable technologies are developing. There is the rise in the careers of the renewable as well as the nonrenewable energy industries. Though, there is much dissimilarity in between the two sectors.

Each of them involves the advantages and the challenges. And, on the other hand, these are related to the unique technologies that play an important character in the current system of energy. For a variety of reasons, from the restricted quantity of fossil fuels that are available to their effects on the environment, there is an increase in the interest in using renewable forms of energy and developing technologies to upsurge their productivity. This rise in the industry calls for a new staff.

7.3. NON-RENEWABLE RESOURCES OF ENERGY

7.3.1. Coal

Coal is a black or brownish rock. The coal is burned in order to produce energy. The ranking of the coal is depending on the rate of carbonization it has gone through. Carbonization is the procedure that the early organisms undergo to become coal. Near about three meters, that is ten feet of solid vegetation crumpled together into one foot of coal (Figure 7.3).

Figure 7.3: The coal is mined and burned for the energy from a very long time.

Sources: https://www.maxpixel.net/static/photo/1x/Coal-Extraction-Coal-Mining-Mine-Excavator-Machine-2781686.jpg

Among all the ranks of the coal, peat is the lowest rank of coal. It has gone through the smallest amount of carbonization. It is a most important fuel in many areas of the world. These areas comprise of Scotland, Ireland, and Finland.

On the other hand, Anthracite is the highest rank of coal among all. Anthracite is produced in regions of the world where there have been massive movements of the earth, such as the creation of mountain ranges. It is to be noted that the Appalachian Mountains are rich in anthracite. The Appalachian Mountains are in the eastern part of the United States.

People all over the world mine coal out of the ground so we can burn it for energy. There are two methods that are used by the people to mine coal. One is underground mining, and the other is surface mining.

Underground mining is used when the coal is placed under the surface of the Earth; sometimes, it is 300 meters that is 1000 feet deep. This is deeper as compared to most of the Great Lakes. Mineworkers or coal miners take an elevator down a mineshaft.

These coal miner function on the heavy machinery. The machinery that cuts the coal out of the Earth and brings it directly above the ground. This can be an unsafe work. This is because of the reason that the cutting coal can emit gases that are really very risky. These gases can have many hazardous effects. The gases can cause explosions and also, it can make difficulties for miners to breathe.

On the other hand, the other method of mining the coal is Surface mining. Surface mining is used when the coal is situated very nearby the surface of the earth. The companies are required to clear the area in order to get to the coal. The companies also take away the trees and soil. After this, the coal can be cut out of the ground in a very easy way. All through this procedure of coal mining, the complete environment is devastated.

Near about half of the electricity in the United States comes from the coal. Coal is of great importance as it helps in many things. Coal helps in delivering power to the lights, refrigerators, dishwashers, and most other things that the people plugin. At the time when the coal is burned, it leaves behind "byproducts." These byproducts are also valuable. The byproducts are used to make many things which the requirements of the people are. The people use the byproducts to make cement, plastics, roads, and several numbers of other things.

7.3.1.1. Advantages and Disadvantages of Coal

Coal is a dependable source of energy. The people can depend on it every day and night, summer, and winter, sunshine or rain. Coal provides fuel and electricity.

The use of coal is also damaging. Mining is one of the most unsafe works all over the world. Coal miners or the mineworkers are exposed to toxic dust. These workers face the hazards of cave-ins and explosions at their workplace.

At the time when the coal is burned, it emits many poisonous gases and pollutants or impurities into the air. The mining for coal can also cause the ground to cave in and make underground fires that burn for years at a time.

7.3.2. Petroleum

Petroleum is a liquid form of fossil fuel. It is also known as the oil or crude oil.

Petroleum is surrounded by the underground creations of rock. In some of the places, oil is bubbled right out of the ground. At the LaBrea Tar Pits, in Los Angeles, California, big ponds of dense oil bubble up through the ground. The leftovers of the animals that got stuck there around thousands of years back are still conserved in the tar. It is to be noted that the majority of the oil from the world is still found deep below the ground. This is drilled through the earth to have access to the oil. Some deposits are located on land, while the others are beneath the ocean floor (Figure 7.4).

Figures 7.4: Petroleum is extracted from the ocean floors.

Sources: https://c.pxhere.com/photos/4c/71/cranes_industriehafen_wismar-715058.jpg!d

As soon as the oil companies start boring with the help of "drill rig," these companies of oil can extract petroleum 24 hours a day, 7 days a week, 365 days a year that is at every time of the day and night and throughout the year. There are many sites of the oil that is successful in making oil for near about 30 years. At sometimes even these oil sites can produce oil for much longer time.

When oil is below the ocean floor, then the companies drill offshore. These companies must build an oil platform. Oil platforms are some of the biggest structures in the world that are made by the man.

The oil must be refined as soon as the oil has been drilled. Oil consists of many chemicals other than the carbon. And also, the process of refining the oil takes some of these chemicals out from it. The people use oil for several numbers of things. Near about half of the petroleum of the world is transformed into gasoline. The rest, which is left, can be processed and also, it can be used in several numbers of things.

It can be used in liquid products such as nail polish and rubbing alcohol, or also, it can be used in the solid products such as water pipes, shoes, crayons, roofing, vitamin capsules, and thousands of other items.

7.3.2.1. Advantages and Disadvantages

There are many benefits to drilling for oil. It is low-cost to extract as compared to the other fossil fuels. Petroleum is also a dependable and reliable source of energy and money for the resident community.

Oil provides with thousands of amenities to the people all over the world. In the form of gasoline. It is to be noted that it is a portable source of energy that gives the power to drive places. Petroleum is also an ingredient in many of the items that the people rest on.

Though, the burn of gasoline is also very dangerous to the environment. It emits hazardous gases and fumes into the air that everyone breathes. There are also the chances of an oil spill. And also, if there is any kind of issue with the drilling machinery, then the oil can burst out of the well and spill into the ocean or the neighboring land. Oil spills are environmental tragedies, especially offshore spills. Generally, oil floats on water, so it seems like a food to fish and ruin birds' feathers.

7.3.3. Natural Gas

Natural gas is the other type of fossil fuel. Natural gas is trapped underground in reservoirs. It is mostly made up of methane. Many people may have smelled the methane before. The decaying material in landfills also emits methane. And, methane smells like rotten eggs.

There is so much natural gas is present underground. This is the reason that the natural gas is calculated in million, billion, or trillion cubic meters.

Natural gas is located in the deposits that are a few hundred meters underground. All the companies drill straight down in order to get the natural gas out from the ground. This is how the companies get natural gas for the use.

Though, natural gas does not form in big open pockets. Natural gas is stuck in rock formations that can stretch for kilometers. In order to get to the natural gas, some of the companies use a process. This procedure is known as "hydraulic fracturing," or it can also be known as fracking. Hydraulic means the companies make use of water, and fracturing, on the other hand, means to "split apart."

The whole procedure of reaching out to the natural gas makes use of high-pressure water to split apart the rocks underground. This emits the natural gas that is stuck in rock formations. If the rock is too hard, the companies can send acid down the well in order to soften the rock. They can also make use of tiny grains of glass, or the sand can also be used to prop open the rock and let the gas discharge from it (Figure 7.5).

Figure 7.5: Natural gas can be used for many purposes like heating, cooking, in automation.

Sources: https://www.maxpixel.net/static/photo/1x/Energy-Natural-Gas-Gas-Flame-Flame-Oil-Drillers-2720980.jpg

The natural gas is used for the purpose of heating and cooking. Natural gas can also be burned in order to produce electricity. It is useful in several numbers of ways. People all over the world depend on natural gas to give power to lights, televisions, air conditioners, and kitchen appliances in their households.

It is to be noted that the natural gas can also be converted into a liquid form, and this liquid form is known as liquid natural gas (LNG). LNG is much cleaner as compared to any other fossil fuels.

The LNG takes up a smaller amount of space as compared to the gaseous form. The quantity of natural gas that would fit into a big beach ball would fit into a ping-pong ball in the form of liquid and not gaseous. LNG can be easily stored, and it can be further used for different purposes. LNG can even be a substitution for gasoline.

7.4. OTHER NON-RENEWABLE RESOURCES OF ENERGY

Fossil fuels are the top non-renewable energy sources all over the world. However, there are many other alternate non-renewable resources of energy. There are some other non-renewable resources of energy that provide the energy but are not beneficial for the environment.

7.4.1. Nuclear Energy

Generally, the nuclear energy is considered as another non-renewable energy source. Even though nuclear energy in itself is a renewable energy source but on the other hand, the material that is used in nuclear power plants is not (Figure 7.6).

Figure 7.6: Nuclear energy is one of the non-renewable alternate sources of energy.

Source: https://upload.wikimedia.org/wikipedia/commons/4/4e/Nuclear_Power_Plant_Cattenom.jpg

Nuclear energy produces the powerful energy in the nucleus, or core, of an atom. Nuclear energy is emitted through the process of nuclear fission. Nuclear fission is the process wherein the nucleus of an atom splits. Nuclear power plants are the complex machines that can have control on the nuclear fission in order to generate produce electricity.

The material that is used most often in the nuclear power plants is the element uranium. Even though, the element uranium is originated in rocks all over the world. Generally, the nuclear power plants make use of very rare category of uranium, U-235. Uranium is a non-renewable resource of energy.

Nuclear energy is a famous method of producing the electricity all over the world. Nuclear power plants do not contaminate the air or releases the greenhouse gases (GHG). Nuclear power plants can be built in rural or urban areas, and do not abolish the surroundings around them.

Though, nuclear energy is hard to harvest. Nuclear power plants are very complex to build and run. There are many communities that do not have the scientists and engineers to develop a harmless and dependable nuclear energy program.

Nuclear energy also makes the radioactive material. Radioactive waste can be very toxic. The radioactive waste can cause burns and can increase the risk for cancers, blood diseases, and bone decay among the people who are exposed to it.

7.4.2. Biomass Energy

Biomass energy is a renewable energy source. It can also be a non-renewable energy source. Biomass energy uses the energy that is originated in plants (Figure 7.7).

Biomass energy depends on biomass feedstocks. Biomass feedstocks are the plants that are processed and burned in order to make electricity. Biomass feedstocks can also involve crops such as corn or soy, as well as wood. If the people do not replant biomass feedstocks as quick as they make use of them, then the biomass energy, on the other hand, becomes a non-renewable energy source.

Figure 7.7: Biomass is being used more often as a source of energy.

Source: https://s0.geograph.org.uk/geophotos/02/99/35/2993546_c2ca0051.jpg

7.5. SOCIAL IMPACT OF NON-RENEWABLE RESOURCES OF ENERGY

The major part of the energy is drained out from the Earth. This energy is drained out from the sources of fuel that are present in the form of fuels. These resources are the non-renewable resources of energy. The dependence of the modern society has increased exponentially on the mineral and fossil fuel, which are mainly derived from the non-renewable energy sources.

There is a difference in the availability, cost of production, and geological and economical distribution of the non-renewable resources of energy. Even though the solid fuels, the energy resources of the fossil fuels and the mineral commodities are non-renewable in nature (Figure 7.8).

Figure 7.8: There are many social impacts of the non-renewable resources of energy.

Source: https://upload.wikimedia.org/wikipedia/commons/thumb/d/dd/Protest_against_Dakota_Access_and_Keystone_XL_Pipelines_20170126-1638.jpg/1024px-Protest_against_Dakota_Access_and_Keystone_XL_Pipelines_20170126-1638.jpg

Still, the metals that are extracted from these sources have the capability to be used again and again. But there is a limit for the recycled use of the energy that is extracted from these sources of energy.

Although, there are many studies that have indicated that the stock of the secondary resources of the non-renewable resources of energy has depleted in past few years. On the other hand, the use of these non-renewable resources has increased. Still, the world needs some alternate to this kind of resources, and similar need will be displayed in the future. There has been a growth in demands for the energy through these resources. This has already begun to restrict the availability of the non-renewable resources in the crust of the earth.

The Earth is not facing any depletion of the important mineral resources, at least in the near future. But, the ability to explore as well as to extract the energy from the non-renewable resources of energy is being restricted in many different countries. This has been done by competing land use, as well as by adopting the major steps on the political and environmental issues.

The extraction of the natural resources has the need of a clear aim on the sustainable development. This involves the economic, environmental, social, and cultural aspects. There is no surety about the most important resource for the energy in next 100 years. But this is also very sure that the society will still require the energy and a wide range of raw materials for the sustainable development.

It is expected that the sources that might be required by the society for the sustainable development are oil and gas, coal. It might also include nuclear material like uranium, thorium, geothermal, and metallic minerals. Apart from the nuclear material, the industrial and specialty minerals, such as cement, raw materials, and rare-earth elements might also make to the list of requisite materials.

Apart from amassing the materials for the sustainable development, an approach which can be used throughout the world for examining the amplitude and the future availability of such resources is required. This would be an approach that, with appropriate international collaboration, was started within the triennium of the International Year of Planet Earth.

There are many mineral resource examinations have been started throughout the world. These methods include the collaborations between the governments of the different countries, which have already taken up the examination methods and the other set of countries where these methods have not yet been initiated. By marking the year as The International Year of Planet Earth has helped the people throughout the world to keep their focus on how the proper usage of the geosciences can be used to generate prosperity at local as well as global level. Apart from this, these methods have also helped to address the sustainability issues in both developed and developing countries.

The conventional energy sources that are based on oil, coal, and natural gas have a damaging impact on the economic progress, environmental conditions, and the human life. The uses of the traditional fossil fuel-based energy resources have been facing the opposition and an alleviating pressure on some of the environmental fronts. Perhaps, the most critical challenge that the non-renewable resources are facing is the future usage of coal after the Kyoto Protocol targets the reduction of the greenhouse gas at a global level. The increased use of the renewable energy sources has led to the development of the current supply that matches to somewhere around to 15% and 20% of total energy demand of the world. The supply of the energy is dominated by conventional biomass. This conventional or traditional biomass mainly includes fuelwood that is used for cooking and heating. This trend of usage is especially observed in the developing countries of Africa, Asia, and Latin America. Many studies have revealed that the modern renewable energy sources are only about 2% in the contribution of the total energy demands in these countries. These modern renewable energy resources include solar energy, wind energy, modern bio-energy, geothermal energy, and small hydropower.

There are many cases and observations that have examined the potential contribution of the renewable energy sources to the energy supplies of the world. These studies indicate that in the second half of the 21st century, the contribution of the renewable sources of energy might range from the present number of near about 20% to more than 50% with the required policies in correct place. The potential of renewable energy sources is enormous as they can in principle meet many times the world's energy demand.

7.6. ENVIRONMENTAL IMPACTS OF NON-RENEWABLE RESOURCES OF ENERGY

There are three major fossil fuels that are being used and are being discussed in the global scenario. The three major fossil fuels are oil, coal, and natural gas. Each of these is valuable source for the energy and has played a significant role in the last few decades. Each of the sources has their own advantages and disadvantages.

The simplest fact that increases the usage of the fossil fuels is that the fossil fuels are easy to use. They are easy to transport, and there are many countries across the world that have the specific infrastructure that has been specially designed for the fossil fuels. Moreover, the fossil fuels are inexpensive in usage.

Unfortunately, it is a very well-known fact is that the fossil fuels are not sustainable and do not support the sustainable development. The rate of the usage of the fossil fuels is much higher than the rate of the creation of fossil fuel. The excessive usage or the burning of the fossil fuels is the main cause of the majority of the environmental impact. The main impact that has been created by the burning of the fossil fuels is that they produce GHG that have an adverse effect on the ecosystems.

The oil is the main type of fossil fuel that has the major share in the adverse effect on the environment. The environmental impact of the burning of the oil is well-known. When the oil is burnt or used in the factories, cars, and other modes of automation, it releases many harmful or toxic chemicals into the air (Figure 7.9).

Figure 7.9: Non-renewable resources have heavily impacted the environment.

Sources: https://upload.wikimedia.org/wikipedia/commons/7/7c/Gibraltar_East_Side_Water_Catchments_in_1992.jpg

Along with this, there is a constant risk of the oil spills, which have seen a rise in the past few decades. Oil spills are very harmful, especially for the aquatic ecosystem. The increased rate of the oil spills has led to the serious issues in the marine and aquatic ecosystem throughout the world. In addition to this, the oil needs to be refined. The refining of the oil and the disposal of the unwanted components is a source for another toxic waste.

It is very troublesome to find and extract the coal from the environment. The mining of coal has led to the landslides in the mountaintops and hillsides. The process of extraction of coal also creates the waste that can inflict damage to the environment and pollute the groundwater table inside the Earth. The contaminants that are released from burning the coal have been the cause for the large range of environmental issues, which mainly includes the acid rain.

Now, the effect of natural gas on the environment is very less than as compared to that of coal and oil. This is the main reason for the availability of the numerous proponents of natural gas. But there are adverse effects of the natural gas as well. The use of natural gas emits many GHG into the atmosphere. Moreover, the transporting of the natural gas can turn up to be expensive. The main reason for the aggravated cost of them because there is a need to liquefy the natural gas before it can be exported for the usage in households, vehicles.

7.6.1. Greenhouse Gas Emissions

The major consequence of the usage of the non-renewable sources of energy is the emission of the GHG. The main GHG that are being an adverse effect to the environment are carbon dioxide and methane. The emission of these two gases has led to the extreme climate changes in last few years (Figure 7.10).

Figure 7.10: Greenhouse gas emissions are responsible for global warming.

Source: https://climate.nasa.gov/system/news_items/main_images/2412_air-pollution-smoke-rising-from-plant-tower768.jpg

Different types of non-renewable energy fuels emit different levels of GHG.

Coal is measured as the worst emitter of carbon dioxide amongst the non-renewable resources of energy. For example, the emissions of the carbon dioxide from the electric power sector that was calculated in the year 2015 indicates that near about 71% of these emissions were characterized from coal. On the other hand, the emissions of carbon dioxide from the natural gas account only about 28% of the total emissions in the year 2015.

This is true to the fact that the natural gas emits a very less amount of carbon dioxide. If compared with coal, the natural gas specifically emits near about 50% to 60% off the carbon dioxide as that from coal. Moreover, it emits around 15% to 20% of the heat-trapping gases as compared to the gasoline, which is a form of oil, when used to power a vehicle.

However, these results do not imply that the natural gas can help in solving the issue of climate change. This is because of the methods that are being used in the extraction of the natural gas. The methods of drilling and extracting the natural gas from the reservoirs results in the emission of methane. Methane has much more potential than many of the GHG. Methane is around thirty-four times more capable of trapping the heat than the carbon dioxide.

The concerns about the emission of the greenhouse gas and the alterations in the climate are cross-cutting. The issues are not just about the direct effects of the increasing temperatures and altering weather patterns. There are many alterations, such as floods or dry seasons that heavily affect the human livelihoods.

The frequent changes in the climate is affecting the ecosystem and decreasing the capacity of the ecosystems to adapt to the ever-changing conditions. Moreover, such frequent changes are threatening the entire biodiversity and the important ecosystem services on which the human lives are dependent.

According to many studies, the burning of the fossil fuels has added to near about three-fourths of the total emissions in the last twenty years that are leading to the global warming. According to many institutions of petroleum, coal, and natural gas, the fossil fuels are responsible for near about 42%, 32%, and 27% of greenhouse gas emissions in the United States in the year 2014.

7.6.2. Air Pollution

The impact of the non-renewable energy sources is not only affecting the terrestrial environment but also contaminating the atmosphere. The main harm to the atmosphere is done by the alleviated rate of the amount of greenhouse gas emissions. The excessive usage of the non-renewable resources of energy also emits many different types of pollutants. These pollutants have an adverse effect on the health of the people and the environment.

For instance, there are power plants that use coal as the main fuel for the generation of energy. These power plants are the single largest source of emissions containing emissions in many developed countries across the world, especially in the United States of America.

When the emissions containing mercury are released into the air, the mercury settles down on the deposits on the ground or in water. Through this way, the mercury can get deposited in the natural resources and enter the food chain. As soon as the mercury enters in the food chain, it enters in the organisms of species that are the inhabitant of those specific areas (Figure 7.11).

Figure 7.11: Air pollution has led to destruction of environment.

Source: https://cdn.pixabay.com/photo/2016/10/19/12/38/industry-1752876_960_720.png

The air pollution has a reflective effect on the biodiversity in the world. On the other hand, the air pollution also creates the real problems for the human being. For Example, many studies that have been conducted across the world have discovered that the exposure to mercury can lead to many neurological and neuro-behavioral disorders in the embryos and young children.

Other air pollutants emitted due to fossil fuel combustion include sulfur dioxide, nitrogen oxides, and particulate matter.

7.6.3. Acid Rain and Water Pollution

The adverse effect of the non-renewable sources of energy is not just limited to air pollution or land pollution. The pollutants which are emitted in the atmosphere enter the water cycle by mixing with the water content and are very harmful in this way. The heavy pollutants that are released into the air become a part of the water cycle.

This leads to the formation of the acidic base in the clouds and thus leads to acid rain. The case of acid rain is a very harmful effect of the pollutants that are released from the use of the non-renewable sources of energy. The chemicals like the sulfur and other chemicals are emitted in the atmosphere and are introduced into the process of cloud formation from industrial processes. Chemicals suspended as the base particles in the air covert the rainwater mildly acidic (Figure 7.12).

Figure 7.12: Acid rain has led to destruction of many habitats across the world.

Source: https://upload.wikimedia.org/wikipedia/commons/0/0c/Acid_rain_woods1.JPG

Acid rain is very harmful for living and non-living things in the biodiversity. The acid rain can cause corrosion to the equipment and can cause a problem in the proper functioning of the local ecosystems. As far as the environment is concerned, the acid rain alters the acidity of the water present in the natural water bodies like the lakes and streams.

This turns out to be fatal for the fish and other aquatic organisms that dwell in those natural water bodies. Along with this, the acid rain damages the trees in the forests across the world, thereby making the forest ecosystem very weak and hollow.

Many countries across the world conducted some studies around two decades ago. These studies have stated that near 5% of lakes in the region of New England have already turned acidic and almost 2% of the natural water bodies in the same region could no longer support trout.

There is one more feature of water pollution that has been increased because of the extreme usage of the non-renewable sources such as coal, or nuclear energy. This new feature of the pollution is called the thermal pollution.

Both fossil fuel plants and nuclear power plants needs water to run and thus, helps in cooling the power plant. The water that is used by these plants are also used for several numbers of use is also needed for several other purposes, such as maintaining local ecosystem functions or agriculture. After the usage of the water, the plants release the water back into the environment. This process changes the temperature of the ecosystem. Therefore, the quality of the water in the ecosystem degrades.

The re-heated water is again introduced to the environment. This re-heated water contains very low levels of the dissolved oxygen. This leads to the downgrading of the native wildlife by stress on the natural resources. There are many happenings that are a consequence of the thermal pollution. For example, an alleviated heart rate of the fish or other aquatic animals and decreasing fertility, among others.

7.6.4. Land Pollution and Waste Generation

The extraction and the drilling process of the non-renewable resources of energy also contaminate the land. It is also important not to forget environmental impacts that come about as a result of the extraction of non-renewable resources or the disposal of the waste that is generated from the non-renewable resources of energy.

There are many clear evidences that explain the impact of the surface mining. The impact of the surface mining is both in the short and long-term. For examples, there are huge volumes of excess rock or soil that is dumped in other nearby locations such as valleys or plains. This process of dumping the waste from the surface mining affect the ecosystems present in those places.

The extraction process leaves the site of mining barren and useless. The land that is being mined suffers a lot in the long term. These sites are left with poor soil quality and sometimes due to many harmful chemicals that are used on the lands during the process of extraction. Eventually, the land ends up being polluted as well as the nearby water reservoirs get contaminated.

There are many cases of the waste disposal of nuclear energy that has been the main reason of contamination of the land resources. The waste disposal of the nuclear energy is particularly relevant in context with the land pollution. The extraction of the uranium and the waste disposal of the nuclear waste create some very serious issues (Figure 7.13).

Figure 7.13: Land pollution is a result of the improper waste disposal.

Source: https://c.pxhere.com/photos/88/a8/land_fill_trash_dump_environment_fill_pollution_sanitary_litter-622469.jpg!d

These issues have a long-lasting impact on the ecosystem. For the consequences of the waste disposal of the nuclear reactors, there is no long-term solution has been discovered. This is so because of the radioactive nature of the ore that is required for the raw material at the nuclear reactor and the radioactive waste that has to be disposed.

7.6.5. Oil Spills and Other Accidents

Finally, there is something that is needed to be said about unintended consequences or rather unexpected and unintentional effects. Oil spills are very harmful to nearby shores and ecosystems. In order to quantify this, there were some analysts or experts who estimated the British Petroleum's oil spill of the year 2010 would cost 2.5 billion US dollars in losses to the Louisiana fishing market alone. On the other hand, Florida was estimated to lose 3 billion US dollar in tourism income.

There were some ecologists or Biologists who were concerned that a species of algae, which is important to hundreds of species of animals, would be wiped off. This is because of the reason of the oil that was emitted in the Gulf of Mexico. Humans and nature were affected in the same way (Figure 7.14).

Figure 7.14: Oil spill has caused havoc in the aquatic ecosystem.

Source: https://response.restoration.noaa.gov/sites/default/files/images/27/westchester-spill_for-sos-blog.jpg

Disturbing effects were seen on nature as well as the humans. These effects were because of the nuclear disaster in Chernobyl. These were the coincidences which took place because of several numbers of factors. Some of them are the negligence or the carelessness, failure of the technology, absence of preparation, or it was because of the combination of all those.

7.7. ECONOMIC IMPACT OF NON-RENEWABLE RESOURCES OF ENERGY

The New Webster's Dictionary describes the "renewable" as "replaceable naturally or by human activity." The illustrations of renewable resources comprise of the trees and other plants, animal populations, groundwater, etc.

In comparison to it, the non-renewable resources on the other side are those that are not replaceable or replaced so gradually by the natural or artificial procedures that for all practical purposes, as soon as they are used, they would not be available again in the future within any reasonable time frame.

The very clear examples are that of oil and mineral deposits. There were many newspapers or reports that have added another dimension to this difference between the renewable and the nonrenewable resources. Since, the money matters are related with the allocation of resources which are very rare, for an economist non-renewable resources not only have a fixed stock, but on the other hand, they are also in a restricted supply as compared to the demand for them.

Therefore, the long-standing growth trees with life spans of as much as 1,000 years. On the other hand, the renewable by the common definition, may be classified as non-renewable by economists. This is because of their relatively slow progress to maturity and scarce remaining stands. It is to be noted that they may also be naturally unique and are not reproducible. In the same way, while coal would be considered as the non-renewable by some of the economists, most resource reports or newspaper would also consider it renewable. This is because of the massive remaining stock.

It has been estimated that at the present rates of consumption of near about one billion tons per year, there is enough coal to last for near about 3,000 years. From an economic viewpoint, there is no instantaneous shortage of coal because of its fixed stock. It is as if it were renewable. Also, there exists no scarcity rent in connection with its extraction. By tradition, the most important economic problems in the study of non-renewable resources have included the foreseeing of the future production and price trajectories, as well as the time of the probable resource exhaustion. In addition to this, there has been an effort to recognize the influence of alternative market structures on the forecast trajectories. These market structures are like pure competition, monopoly, and a cartel-dominated oligopoly. In the present times, the environmental costs that are linked with the extraction and consumption of non-renewable resources have also come into attention.

7.8. FUTURE OF NON-RENEWABLE RESOURCES OF ENERGY

The long-term obtainability of non-renewable sources of energy depends not just on the reserves or resources for primary fuels, but on the other hand, it also depends on their annual extraction from the deposits. If the total reserves of a specific fuel are divided by the quantity that is presently taken out or mined, then this gives the reserves-to-production ratio for the particular fuel.

In context to the situations that were prevailing in the year 2007, conventional crude oil will be available all over the world for forty-two years, natural gas for sixty years, black coal for 129 years and lignite for 286 years.

In consideration to the reserves in the year 2005, uranium has a reserve-to-production ratio of 70 years.

Though, all this demonstrates a snapshot and assumes that the consumption, which is based on the current reserves, is continued at the present level in the upcoming time. What it does not take into consideration is that the progressions in energy-saving technology and the replacement successes lessen the consumption. On the other hand, the new deposits were discovered as a consequence of better-quality exploration technologies. These new deposits further help in up surging the reserves.

From a viewpoint of sustainability, nuclear electricity displays a thought-provoking dilemma. On the one hand, nuclear electricity generates no carbon emissions. This is a main sustainable benefit in a world that is facing anthropogenic change in the climate.

On the other hand, nuclear electricity generates hazardous waste that:

- Must be kept out of the surroundings for thousands of centuries;
- Can generate bomb-grade plutonium and uranium. These elements could be diverted by terrorists or other people in order to abolish cities and poison the environment, and
- Threatens the natural and built surroundings by the leaks of long-lived radiation by chance.

The considerate or thoughtful scientists, policymakers, and citizens must evaluate the advantages of this source of carbon-free electricity against the environmental risk of storing spent fuel, the societal risk of nuclear proliferation, and the impact of accidental or deliberate release of radiation.

There are very less cases of humans that are having the control to transform the dynamics of the earth forever. One of the examples is the global climate change from carbon emissions, and the radioactivity from the explosion of a several numbers of nuclear weapons is another example.

Nuclear electricity touches both of these opportunities in both the positive as well as negative way. In case of positive side, it helps for the lessening of the carbon emissions. On the other hand, on the negative side for the risk of nuclear proliferation.

The advent of the nuclear electricity has originated very quickly on the global level. The advent of the energy all over the world in the energy scene. There were many developments in the nuclear technology that started at the end of World War II. These developments initially started for the military ends. The nuclear energy rapidly gained a new peacetime path for very cheap production of electricity. After eleven years of the end of World War II, the very first commercial nuclear reactor that was able to produce electricity was set up at the Calder Hall in Sellafield, England. This is a very small period when it comes to the terms of the establishment of the electricity producing plant by the use of the nuclear energy.

The number of nuclear devices rises up progressively to more than 400 by the year 1990, 4 years after the Chernobyl disaster in the 1986 and 11 years following Three Mile Island in the year 1979. Since the year 1990, the number of operating reactors has continued to be approximately flat, with new construction which is balancing the retiring because of the public and government unwillingness to continue further with nuclear electricity expansion plans.

The result of this discussion will determine whether the world experiences a nuclear renaissance that has been in the making for several years. The global discussion has been strongly impacted by the unlikely nuclear accident in Fukushima, Japan, in March 2011.

The Fukushima nuclear disaster was produced by an earthquake and tsunami. These calamities inactivated the cooling system for a nuclear energy complex which comprises of operating nuclear reactors and storage pools. These storage pools are used for the storage of the spent nuclear fuel under water. Ultimately, this leads to a partial meltdown of some parts of the reactor cores and emits a considerable amount of radiation.

This happening, twenty-five years after Chernobyl, reminds the people all over the world that safety and public confidence are especially important in nuclear energy; without them, expansion of nuclear energy will not happen.

REFERENCES

1. Akella, A., Saini, R., & Sharma, M., (2008). *Social, Economic and Environmental Impacts of Renewable Energy Systems*. [online] Citeseerx.ist.psu.edu. Available at: http://citeseerx.ist.psu.edu/viewdoc/download?doi=10.1.1.461.4230&rep=rep1&type=pdf (Accessed on 21 May 2019).
2. Fisher, M., (2019). *11.2 Non-Renewable Energy Sources*. [online] Openoregon.pressbooks.pub. Available at: https://openoregon.pressbooks.pub/envirobiology/chapter/11-2-non-renewable-energy-sources/ (Accessed on 21 May 2019).
3. Friedrich, W. W., & Sinding, L. R., (2012). *Non-Renewable Resource Issues*. [online] Available at: https://www.researchgate.net/publication/281045111_Non-Renewable_Resource_Issues (Accessed on 21 May 2019).
4. Kinhal, V., (2019). *Non-Renewable Resources*. [online] Greenliving.lovetoknow.com. Available at: https://greenliving.lovetoknow.com/Non_Renewable_Resources (Accessed on 21 May 2019).
5. Morse, E., & Shea, J., (2019). *Non-Renewable Energy*. [online] National geographic society. Available at: https://www.nationalgeographic.org/encyclopedia/non-renewable-energy/ (Accessed on 21 May 2019).
6. Smith, S., (2006). *Renewable and Nonrenewable Resources*. [online] Penn state extension. Available at: https://extension.psu.edu/renewable-and-nonrenewable-resources (Accessed on 21 May 2019).
7. Vonk, A., (2008). *Non-Renewable Energy Sources*. [online] Science learning hub. Available at: https://www.sciencelearn.org.nz/resources/1570-non-renewable-energy-sources (Accessed on 21 May 2019).

8

Technical Advancements and Renewable Resources

CONTENTS

This chapter talks about the technical advancements and renewable resources with the help of different schemes and examples.

This chapter also talks about the subdivision of a decade of change in the innovation and distributed energy resource (DER) systems, technological advancements, and their impact on renewable resources.

This chapter also discusses about grid storage technologies as well as the greatest clean energy advance, the evolving policy landscape, and promising future for renewable energy.

8.1. INTRODUCTION

The eagerness for renewable forms of power has taken something of a rough treatment in present years, with government subsidy being taken back and saver emotion going cold. The extensively held estimation is that there wants to be a gesticulate of technical advances in the renewable energy division for the production to get back on path.

From solar farms to aground and offshore wind, governments in many developed countries, have scale back many of their more determined renewable energy projects, in favor of more reliable if more pollute fossil fuels.

Though, there is a harsh need for sustainable sources of energy to come into view over the coming decade, not least in the rising parts of the planet that lack the natural resources originate somewhere else. All through Africa, there is a distressed need for energy to power the economy of many countries that have been in a weak position by overly luxurious oil and gas import and a lack of contemporary infrastructure to use them professionally.

Due to the characteristic high price that comes with creating technological advances in renewable energy, there has been small development by the industry in the continent. Things look to be altering; however, with a figure of investments in current years that could see renewable energy take off from corner to corner Africa.

Nowadays, renewable energy responsible for close to 19% of the whole energy consumed across the globe from origins such as hydro, solar, wind, geothermal, and biomass energy. Renewable energy is rapidly gathering impetus due to technological advancements as well as greater than before demand for energy that is reasonably priced and green.

As renewable energy market and industries full-grown, they are creation way for a broad range of opportunity, with new sets of challenge. Large-scale acceptance of dispersed energy systems and automation technology, along with advancement in storage technology, is accelerating. The integration of renewable energy into major stream energy production is also increasing.

DER systems are in advance popularity because they can decrease or potentially eliminate costs, inefficiencies, and interdependencies linked with traditional central energy systems. With advancement such as improved battery chemistry and efficiency, mechanization to distantly keep an eye on and control battery installation has compulsory usefulness companies and grid operator to rethink how demand organization must be optimized.

Automating energy substations allow data to be second-hand from intelligent electronic devices (IED) to supervise, control, and mechanize substation operation. This has resulted in greater than before reliability of both the DERs as well as conservative grid over a broad range of operating circumstances. Supervision of power release systems in substations reduce the incidence of outages and can allow for quick reallocations of energy capital using isolated manages mechanism.

The development of renewable energy over the past decade has surpassed all prospects. Global installation capacity and manufacture from all renewable technologies have greater than before substantially, and behind policies have continued to increase to more countries in all regions of the planet.

Development in the early 2000s show an upward trend in global renewable energy speculation, capacity, and integration across all division, yet most usual projection did not forecast the strange growth of renewables that was to open out over the decade ahead. Several factors set the base for this quick growth. The variety of energy crises—beginning in the 1970s—and the following economic downturn underscore the role of energy for both nationwide and financial security. At the similar time, a handful of ground-breaking countries such as Germany, Denmark, Spain, and the United States has created dangerous markets for renewable, which flock early technological advance and economy of scale, background the stage and serving to fuel the past decade of likely to explode market growth. Growing stress on mitigating climate change and adapt to its impact has further contributed to the impetus (Figure 8.1).

Figure 8.1: Solar water heater panels.

Source: https://c.pxhere.com/photos/ac/dd/water_heater_solar_sun_energy_ecology-1331583.jpg!d

Renewable energy's payment to the global heat, power, and convey sectors has greater than before steadily. While growth in renewable share of total power use has been moderate by increase in both population and world energy requirement—most notably in rising and emerging economy—renewable energy markets and technology development have accelerated rapidly, even when compare to other rapidly rising technologies such as mobile phones.

Although the previous decade has seen marvelous advances in the electrical energy sector, the renewable heat and cooling division has lagged behind. This is in spite of the marked growth since the year 2004 in the use of geothermal, solar thermal, and biomass heat technologies for water and space heating, process heat, and cooling.

Given that the split of heating and cooling in final power require is much bigger than that of electricity, development growth of renewable energy in this division is crucial. In the convey sector, the use of renewable energy in the shape of bio-fuels grow at a rapid pace for much of the precedent decade. Biodiesel production greater than before twelve-fold, ethanol manufacture, already at a higher preliminary point, grew three-fold. Over a similar period, a small but rising use of gaseous biofuels in convey emerged, as well as initiative to link renewable energy with electric transport.

8.2. DECADE OF CHANGE

Global perception of renewable energy has shifted significantly. Ten years ago, people extensively acknowledged the possibility of renewable energy, but large-scale use still had to be established. Now ten years on, ongoing technology advances and fast deployment of much renewable energy technology—particularly in the electricity division—have amply established their impending.

Nowadays, renewable energy technology are view not only as gear for civilizing energy security and extenuating and adapt to climate change, but are also more and more recognized as savings that can provide direct and indirect financial compensation by plummeting dependence on import fuels; improving local air quality and safety; advance energy right of entry and security; propel economic development; and creating employment (Figure 8.2).

Figure 8.2: Solar park–Wind park renewable.

Source: https://c.pxhere.com/photos/4b/01/east_frisia_cows_windr_der_graze_bushes_agriculture_dairy_farming_livestock-636740.jpg!d

Moribund costs have also played an important role in the growth of renewable energy use in current years. Several renewable energy technologies are today cost-competitive with conservative generation technologies, even before the surroundings and other externalities are taken into deliberation.

Strange growth in renewable energy markets and their worldwide spread has also led to an important rise in the number of manufacturers, the scale of industrialized, an overall add to in number of jobs install and servicing renewable energy technology, as well as growth into new markets. This is mainly true for the solar PV and wind power industry, in spite of experience industry consolidation, ambitious by decreasing costs.

Ten years ago, use and developer of renewable energy were intense in Europe, the United States, and Japan. Their early lead in the renewable energy worldwide markets cemented the way for them since then, markets, developed, and asset have expanded to another region. China has turn out to be the world leader in renewables manufacturing and install capacity, have increased asset in the sector nearly every year over the preceding decade.

Increasing amount of money are now fluid to rising and emerging countries across Africa, Asia, Latin America, and the Middle East, in reply to fast growth in energy demand and a rising interest in renewables. Overseas direct investment in renewable energy and the mobilization of private assets in up-and-coming economies have also contribute to enlargement cross both technologies and regions technologies advances and market growth through early asset in expertise and in strategy design.

Over the precedent decade, the split of people who be short of access to contemporary energy services has fallen by nearly 10% points, down from almost 25%, even as the worldwide population has long-drawn-out significantly.

Renewable have played a role in this development. However, these advances are not consistently spread geologically; large areas of Africa are still without right of entry to modern energy services. Renewable are exclusively positioned to provide needed energy services in a sustainable manner, more rapidly and generally at lower price than their alternatives.

8.3. DISTRIBUTED ENERGY RESOURCE (DER) SYSTEMS

Decentralized, modular energy capital is located in nearness of the cargo they serve, and characteristically use renewable energy resources. Previous to DERs, a key obstacle for usefulness scale growth of solar, wind, and hydro energy was the intermittency linked with this source. With micro-grids (DERs that function in similar with the usefulness grid), fluctuations due to intermittency from these renewable energy source and its collision to the main grid are being mitigate (Figure 8.3).

Figure 8.3: Solar dish.

Source: https://upload.wikimedia.org/wikipedia/commons/2/24/Solar_dish_at_Ben-Gurion_National_Solar_Energy_Center_in_Israel.jpg

For the first decade of the 21st century, Usefulness Company viewed DERs as a contestant – or more like a danger. But over the last couple of years, this viewpoint has underway to change. With greater than before acceptance of solar energy, dispersed energy generation, and storage space, there has been marvelous pressure on usefulness companies to bout face their move towards.

According to a recent survey by famous researchers, about 74% of usefulness companies are helpful of customers installing a behind-the-meter dispersed-grid infrastructure (DERs and energy storage). Usefulness companies are preliminary to understand the reimbursement of the incremental capital and operational charge that they can alleviate with demand decrease during peak-usage period.

Utility company also rises to advantage from not have to buy power from other provider when the tariff is at its acme, driven by great demand. By integrate dispersed energy capital into a utility companies bring in and insist model; these company are able to get smaller the delta flanked by off-peak and peak stress. The advantages from DERs can be classified into four main categories.

8.3.1. Reliability

A 2013 White House report of United States estimates the price rises-adjusted cost of climate-related outages at $25 to $70 billion yearly. With more and more business unable to have enough money downtime when it come to their data centers, member of staff serving at table, developed facilities, and so on, the usual model of central energy grids is being challenged.

Businesses, as well as usefulness companies, can understand the value of having a cross model of DERs, in adding to a central grid, to make sure the uppermost level of dependability from the network. This cross model offers supreme system resiliency cause by power interruption due to faults and failure in the middle grid. Yearly capacity preparation withhold up from DERs throughout peak insists cycle can greatly decrease the load on the major grid and ensure ease of use of high-quality energy.

8.3.2. Efficiency

The United States. Energy Information Administration (EIA) estimate that electrical energy transmission and sharing wounded standard about 6% of the electrical energy transmit and dispersed annually in the United States. An important piece of these wounded are due to line losses linked with transmission of power from the central grid to heaps that are some miles apart.

DERs can in a straight line get better the efficiencies linked with line losses by surroundings up power sources comparatively close to the weight. Heat generated during power generation can be used to heat water or make steam for employ in near developed plants or other amenities. Competence also can be realized by assessment the weight between DERs and the middle grid throughout peak, and off-peak insist cycles.

A well-intended cross model ensures that only best capital is deploy on both the DER as well the central grid sides of power generation and distribution (Figure 8.4).

Figure 8.4: Solar panels installments underway.

Source: https://media.defense.gov/2010/Jul/16/2000342854/-1/-1/0/100302-F-2428H-102.JPG

8.3.3. Sustainability

It is flattering more and more significant for business to set and achieve their yearly sustainability goals so they can get together current reserve needs with no compromise the requirements of future generation. DERs are as long as an option for business to slot in renewable energy sources as an important portion of their sum energy use combines.

For example, 26% of Wal-Mart's worldwide electricity use comes from renewable energy source from both DERs as well as the network. More and more business is preliminary to include renewable energy and energy competence as part of their long-term plan. These businesses are setting management examples of high-quality environment stewardship, which are made possible by DERs (Figure 8.5).

Figure 8.5: Solar panels.

Source: https://c.pxhere.com/photos/f5/ee/sun_energy_solar_panels-864151.jpg!d

8.3.4. Security

Safety and resiliency of the power source are two high-position key factors when it comes to assignment-critical and zero-downtime project. The yearly 2014 annual review account from Homeland Security's Industrial manage systems Cyber Emergency Response team reported that 32% of the cyber incidents, up-and-coming threats, and vulnerabilities were connected to the energy division.

DERs can assist to speak to the rising concern of the risks linked with a middle grid by enable organization to gain grid self-government capability. Structure power road and rail network to support hybrid structural design with DERs sources all through the grid and have the aptitude to isolate critical application can help to reduce the crash that we see nowadays from cyber-attacks to the middle grid (Figure 8.6).

Figure 8.6: Traditional methods of biomass.

Source: https://upload.wikimedia.org/wikipedia/commons/f/fa/Korea-Gyeongju-Cheomseongdae-02.jpg

8.4. TECHNOLOGICAL ADVANCEMENTS AND THEIR IMPACT ON RENEWABLE ENERGY SECTOR

In adding to ecological issues, growth of renewable energy also responds to a need for diversification of power sources. Though renewable energy is hypothetically unlimited, their possible varies with climatic, physical location, and storage space potential. The compensation and disadvantage of the dissimilar renewable energies are surmounting below.

8.4.1. Solar Power

Solar is a moderately expensive energy basis and, therefore, still immature, though preliminary to take off with the rising prices of fossil fuels. Though it cannot alone replace fossil fuels, it has the benefit of achieving important energy savings, but also a spotless energy that does not create toxic waste and that release no conservatory gases.

In addition, the technology of home solar energy is on the rise and is chiefly effective and established. Solar energy still has some drawback because invest in this clean energy is still very costly. Thus, it is often essential to obtain state subsidy, relations, or bank to be able to go on board on a development scheme of solar energy (Figure 8.7).

Figure 8.7: Solar panels can be used domestically.

Source: https://upload.wikimedia.org/wikipedia/commons/4/43/Jyv%C3%A4skyl%C3%A4_-_solar_panels.jpg

8.4.1.1. Solar Thermal Energy

- **Advantages:** Solar thermal energy produces a high yield and return on investment after she allows having hot water for free. It also produces 50% of the useful heat energy in a home.
- **Disadvantages:** Not only is it a very expensive energy, but the return on investment is quite long (about 10 years) and the lifetime of the panels is limited (20 to 25).

8.4.1.2. Photovoltaic (PV) Solar Energy

- **Advantages:** It is ideal for remote sites or sites that are not connected to a large grid.
- **Disadvantages:** Not only the yield is quite low, but the amount of energy produced by the photovoltaic (PV) panels depends on the climate and geographical location.

8.4.2. Biomass

- **Advantages:** It is an energy that emits little greenhouse gas that can be stored. Particularly concerning wood energy, there is a wide availability of the resource, and the price of firewood is not keeping oil prices.
- **Disadvantages:** It can only have a limited supply because the intensive use of biomass would cause negative impacts on the environment such as deforestation phenomena (when intensive exploitation of wood energy), soil erosions, soil, and water pollution (in case of intensive biofuel production).

8.4.3. Wind Energy

- **Advantages:** It's a totally clean and renewable energy, and its exploitation does not generate waste materials or rejection. In addition, small plants allow electricity to isolated sites, and sites, where the turbines are located, are still usable.
- **Disadvantages:** Performance depends entirely on the wind, and the wind does not always blow when you need it. Furthermore, wind turbines are unsightly for the landscape, and they require some maintenance (cleaning the blades, lubrication …) not to lose their qualities because they are moving mechanical systems.

8.4.4. Geothermal Energy

- **Advantages:** It's an energy that produces no waste, and resource is inexhaustible as the received quality geothermal energy is less than the heat from the center of the Earth.
- **Disadvantages:** Overuse of a deposit results in a temperature drop of the basement, so that the quality of the deposit gradually decrease until it runs out completely. In addition, there may be a competition between the water that is pumped to the heat, and the water is pumped to itself (Figure 8.8).

Figure 8.8: Geothermal energy.

Source: https://upload.wikimedia.org/wikipedia/commons/9/9f/NesjavellirPowerPlant_edit2.jpg

8.4.5. Hydropower

- **Advantages:** It is energy available as streams are dry. In addition, it provides high power and can be stored in reservoirs. It also has great potential, as only 20% of sites are being exploited in the world.
- **Disadvantages:** Its use has ecological impacts in the sense that the dams threaten extinction of terrestrial and aquatic species, but they also deal damage to biodiversity. Moreover, the biggest dams sometimes require the displacement of people, and there is always the risk of dam failure that can cause considerable human and material damage (Figure 8.9).

Figure 8.9: Dam hydropower.

Source: https://cdn.pixabay.com/photo/2017/07/11/08/33/dam-2492809_960_720.jpg

8.5. GRID STORAGE TECHNOLOGIES

Scalable, to nowadays when it is careful the summit of future energy organization systems. With advancement such as better battery chemistry and competence, mechanization to distantly monitor and manage series installation has strained usefulness companies and network operators to about-face how to optimize supply- and demand-side organization. One area with the main impact from power storage is eliminate intermittency associated with solar and wind energy.

As per GTM investigate, the United States has deployed 220 MW of power storage in the year 2015, which is far above use in any preceding year. The usual central grid was intended to provide solely to the supply-side of the equation. This is now set to alter towards both the demand- and supply-side, with the slope of energy storage space solution both before and after the gauge.

8.6. THE BIGGEST CLEAN ENERGY ADVANCE

Clean energy first came into existence in the year of 2016, the Paris climate report went into fruitful labor, the cost of solar installation doomed to fall, the investment in the renewable energy soared a lot, offshore wave finally got underway in the united states of America, and the researchers made a series of technical innovations that promises to make sustainable energy source more and more efficient and reliable as well as cost effective.

The very last one is the input, as the development is still the most affirmative way to keep away from the most impact of climate change. Present commercially produced renewable technology can't walk hand in hand with each other as well as the world's predictive demands. Still, they have scaled up gracefully. The united states alone come up backhanded by the year of 2050, in accordance to methodical examination by the National Renewable Energy Laboratory (NREL).

In the meantime, the United Nations' intergovernmental board on climate change over and done with the world must cut the conserved gas emission by as much as 70% by the mid-century, and to almost zilch by the year of 2100. In order to dodge the global extinction, sinking cities and widespread famine.

So, we need more highly efficient renewable energy sources, cheaper storage, smarter grids, and effective systems for capturing greenhouse gases (GHG).

Here are some of the most promising scientific advances of present times.

8.6.1. Artificial Photosynthesis

One of the important lost parts in the group of renewable energy source is a clean unsolidified fuel that can put back gasoline and other transport fuels. One of the most talented potential is artificial photosynthesis, mimic nature's own technique for convert sunlight, carbon dioxide, and water into fuels (Figure 8.10).

Figure 8.10: A young plant in artificial photosynthesis.

Source: https://farm6.staticflickr.com/5061/5679580299_f96f473e49_b.jpg

There has been slow and steady improvement in the ground in present years. But, Harvard scientists Daniel Nocera and Pamela Silvers, in partnership with their various researchers developed a "bionic leaf" that could imprison and convert 10% of the power in sunlight, a big step onward for the field. It's also about ten times enhanced than the photosynthesis of your average natural plant.

The researchers use mediator made from a cobalt-phosphorus alloy to tear the water into hydrogen and oxygen, and then set particularly engineered microorganisms to work gobble up the carbon dioxide and hydrogen and convert them into fluid fuel.

Others labs have also made notable strides in the efficiency and durability of solar fuel devices in recent months, including Lawrence Berkeley National Laboratory and the Joint Centre for Artificial Photosynthesis. This year the latter lab created a solar-driven device that converted carbon dioxide to format at 10% efficiency levels. Format can be used as an energy source for particular fuel cells.

8.6.2. Solar Thermo-Photovoltaic (PV)

Team from one of the elite engineering institute Massachusetts Institute of

Technology researchers came to a result that the growth of a solar thermo-PV device that could possibly push past the theoretical competence limits of the conservative PV used in solar panels. Those normal solar cells can only absorb energy from a portion of sunlight's color spectrum, mainly the graphics light from violet to red.

But the Massachusetts Institute of Technology researchers added a middle component made up of carbon annotates and nanophotonic crystals that jointly drive sort of like a funnel, collect energy from the sun and directed it into a narrow band of light.

The annotates imprison energy crosswise the entire color range, including in the invisible ultraviolet and infrared wavelengths, changing it all into heat energy. As the head-to-head crystals heat up to high temperatures, around 1,000°C, they re-emit the energy as light, but only in the group that PV cells can imprison and convert.

The researchers propose that an enhanced version of the skill could one day break through the theoretical cap of around 30% competence on conservative solar cells. In theory, at least, solar thermal PV could attain levels above 80%; however that's a long way off, rendering to the scientists. But there's additional critical benefit to this method.

Because the procedure is eventually driven by heat, it could last to function even when the sun ducks behind clouds, plummeting the intermittency that leftovers one of the dangerous drawbacks of solar power. If the device were joined with a thermal storage device that could function at these high temperatures, it could offer incessant solar power in every hour of every week.

8.6.3. Perovskite Solar Cells

Perovskite solar cells are cheaper than the usual types, easy to manufacture and obtain, and very well-organized at fascinating light. A wafer-thin film of the material, a class of hybrid organic and inorganic mixtures with a specific type of crystal construction, can imprisonment as much light as a comparatively thick layer of the silicon used in normal PV.

One of the dangerous challenges, however, has been toughness. The mixes that actually engross solar energy incline to fastly degrade, chiefly in wet and hot circumstances.

But various research groups at universities like Stanford, Los Alamos National Laboratory, and the Swiss Federal Institute of Technology, among other premier institutions, made substantial strides in refining the constancy of perovskite solar cells in present times, publication of notable researches in Nature, Nature Energy, and Science sectors.

Meanwhile, other researchers have prospered at increasing the efficiency of perovskite solar cells and classifying promising new paths for additional advances.

8.6.4. Carbon Storage

Electricity generation is responsible for producing 30% of the nation's carbon dioxide, so seizing those emissions at the basis is crucial to any discount plan. Present situations observe advances for numerous developing methods in capturing carbon in power plants, including carbonate fuel cells, as well as at least some applications of existing technology in the real world. There are also so many various negative examples of this situation.

But most of these methods leave open the query of what to do with the stuff after it's successfully captured. And it's not a small problem. The world produces nearly forty billion metric tons of carbon dioxide yearly.

One method, though, appears more promising than originally supposed: interring carbon dioxide and turning it into stone. Since the year 2012, Reykjavik Energy's CarbFix Project in Iceland has been vaccinating carbon dioxide and water deep subversive, where they respond with the volcanic basalt rocks that are copious in the region.

An examination available in Science in recent years discovered that 95% of the carbon dioxide had mineralized in less than 2 years, much earlier than the hundreds of thousands of years many had forecasted. So far, it also doesn't appear to be sieve-like out GHG, which proposes it could be both cost-effective and safer than current interment methods.

But more and more advancements in the research will be obligatory to observe how efficiently it works in other areas, chiefly counting under the ocean floors, outdoor observers pointed out.

8.6.5. Carbon Dioxide to Ethanol

Another talented option for imprisoned carbon dioxide is, fundamentally, reprocessing it back into practical fuels.

Earlier in previous years, researchers at the United States. Department of Energy's (DOE) Oak Ridge National Laboratory tripped onto a technique for changing it into ethanol, the liquid fuel previously used as a preservative in gasoline. The side developed a substance made from carbon, copper, and nitrogen with a surfaced, which focused the electrochemical responses at the instructions of Nanospikes, rendering to a study published in Chemistry Select a well-known magazine in October.

When voltage was given, the device rehabilitated a mixture of carbon dioxide into ethanol at a high level of competence. The resources were also comparatively cost effective, and the procedure worked at room temperature, both serious advantages for any future commercialization.

"We're taking carbon dioxide, a waste product of combustion, and we're pushing that combustion reaction backward," stated by famous author Adam Rondinone in a news announcement.

In adding to changing captured carbon dioxide, the procedure could be used to store excess energy from wind and solar electricity creation.

Some outdoor researchers, however, are skeptical about the early results and are eagerly waiting to see if other laboratories can confirm the answers.

8.7. THE EVOLVING POLICY LANDSCAPE

The worldwide strategy scenery has mainly driven the outburst of renewable energy technologies by catching the eyeball of investment and producing markets that transported about economies of scale and supported technology innovations of the present as well as future. This, in turn, led to decreasing cost-effectiveness, which eventually fuel sustained development. A trickle of countries, particularly Germany, Denmark, and Spain which lead the way and shaped ground-breaking strategies that drove much of the modification witnessed over the past years.

Presently Germany's promise to the "Energiewende." The change to a maintainable economy solely based on renewable energy and energy competence as well as Denmark's promise to 100% renewable energy by the year 2050, are inspirations many other countries around the globe as well as for the cities to aim high for the upcoming next ten years.

In the meantime, from the year of 2004, the several numbers of nations endorsing various methods or approaches of renewable energy with direct strategy support has been hiked three folds, from the value of 45 to the value of 137, and an ever-cumulative number of emergent and emerging countries is location renewable energy aims and passing support policies. Policy aims have become more and more ambitious, while their attention has long-drawn-out to include heating, cooling, and transport, in addition to electricity. Meanwhile, on the other hand, policy system has constantly enhancing at every developing phase, which includes the utilization of policy instruments characterized on the basis and standard of technology, the development of feed-in policies to premium payments, as well as the dispersal of policy outlines to endorse renewable energy use for heating and cooling.

In consideration of all around the world, renewable energy aims together with feed-in prices have had the main impact on renewable energy market outline. Feed-in policies now are on every landmass, with Jordan, Nigeria, Rwanda, Uganda, the Palestinian Territories, Kazakhstan, and Ecuador being among the most present countries to attain relations with them.

In numerous countries, especially in Europe, mutable renewables have attained high sections of penetration in the power sector very quickly. Current power systems were not intended to deal with such a scenario. In answer to that, policy system that has its attention on market construction are turning out to address requirements in relation to complementary and augmented system suppleness, as well as financial recompense for these facilities.

Policies are also preliminary to answer the requirement for long-drawn-out and developed grid substructure, and progressively, they consist of new tools and skills to provision renewables, such as energy storage and smart grids.

While Europe has been the hub of most such improvements to date, countries in other areas are moving rapidly in this way as their stocks of renewable energy upsurge. Policies that inspire local value formation, such as capacity building also have started to emerge in numerous countries.

In recent times, as renewable energy stocks continue to upsurge, rules that have their attention on obligatory grid joining and importance dispatch are flattering progressively significant.

The previous decade has observed profound alteration on the resident level as well. Ten years ago, the mainstream of local administrations did not reflect the possible role for renewables in their energy stream. Over the past ten years, many of them have become leaders in the progression of renewable energy, particularly in mixture with energy efficiency developments regularly beyond efforts taken by state, local, and nationwide administrations. Various numbers of local governments all over the world have set renewable energy aims and passed fiscal inducements or other rules to stand-in the placement of renewables, derived by the desire to produce local employment, decrease energy cost value, address contamination problems, and loan their sustainability aims. Global organizations devoted to supportive sustainability events in local administrations, such as the Covenant of Mayors and International Council for Local Environmental Initiatives which is Local administrations for Sustainability, have observe their relations skyrocket and their effect spread all around the world.

Around the globe, governments at the community, city, regional, island, and even country stages have started to furnace their own change course of action towards a 100% renewable energy future. They are exposing myths about renewables and proving that 100% renewable energy in close combination with energy competence and upkeep is technically possible, frugally beneficial, and informally wanted.

While the image in much of the biosphere has full-grown progressively cheerful, some countries, particularly in Europe which have abridged renewable energy provision, occasionally retroactively, in the past few decades. This is the circumstance, particularly where power demand has weakened in answer to economic go-slow, and where a congestion of conventional power comes into play.

Till now, however, the European Union is on its way to deal with its member-agreed compulsory target to hike the part of renewables in final energy ingesting to 20% by the year of 2020. Allegedly, three European Union Associate States which includes and not limited to Bulgaria, Estonia, and Sweden, have already arrived at their 2020 aims in the year 2012. Deliberations about making 2030 European Union climate and vigorous aims are under progress.

8.8. A PROMISING FUTURE FOR RENEWABLES

Presently the use of renewable energy innovations to give electricity, heating, and cooling and transportation is now emerged across the world, and present trends stated sustained and maintainable development all over the world. Almost ten years before, renewable energy technologies predominantly occupied an environmental niche, having a strong appeal to those who were interested in moving away from conventional fuels for environmental reasons. Presently, renewables prove that, in adding to their ecological aids, they are also a financial driver, creating employment, in help to expand income watercourses, and inspiring new technological innovations. The main idea of attaining high stocks of renewable energy was fundamental ten years ago. Presently, it is considered possible by many researchers.

The promise to 100% renewable energy in numerous sectors by local, regional, and nationwide administrations all over the world is observer to this. Though, the renewable energy subdivision still expressions numerous tests. Grants for fossil fuels and nuclear power persevere, and last to vastly overshadow monetary inducements for renewables.

Moreover, advances, and asset in renewable energy, as well as developments in energy competence, must last even if the upsurge in global temperature is to be incomplete to 2°C above pre-industrial levels. A rapid decarbonization of the energy sub-division with renewable energy skills is needed to contrivance the climate aims.

The past 10-year has set the wheels of development in work for this change, but a concentrated and continued exertion will be needed to fully attain it. With hiked ambitious aims, innovative strategies, and technological innovations, renewables will go on to exceed prospects and stand-in a cleaner energy future technological innovation in renewable energy are somewhat that firms across the globe are capitalizing heavily in.

Many companies are already convinced that renewable technologies are the future strategy of the energy industry.

Recent research conducted by one of the elite premier educational institution Harvard University has projected that innovations in battery competences could lead to renewable energy skills becoming more efficient and cost-effective. The observation states that by using new batteries, prices can be abridged and more energy can be stowed. Such investigation is helping to branch eagerness in the industry, but isolated firms are also capitalizing in new technology.

Iberdrola, which is a Spanish public multinational electric utility company based in Bilbao, is exploiting considerable quantities into technological innovations in renewable energy. Though it has a leading stage in the onshore wind energy marketplace, it is also observing at offshore wind and maritime technology as a way of attaching sustainable bases of power.

Presently there are an amount of marine technology strategies being tested that are capable of attaching the waves and tides from oceans, altering them into electricity and transporting it to land. They comprise wave and tidal examples off the coast of Santoña in Cantabria, Spain, and the Orkney

Islands in Scotland. Iberdrola Company plans to use this skill as part of the world's first tidal turbine collection in the Complete of Islay in Scotland.

Offshore wind skill, in particular, is somewhat Iberdrola is excited about traveling and capitalizing. January in the year of 2014 obvious a significant attainment for Iberdrola as first control from the West of Duddon Sands offshore wind farm was created. The project is being constructed in the Irish Sea and is Iberdrola's first entry into the offshore wind market, with excellent progress being made to date.

The 398 MW project is at the front of industry labors to decrease costs across the division. Iberdrola's asset in a 50 million euro modified offshore wind terminal at Belfast harbor – with its fitting and pre-assembly capabilities has also paved the way for the construction of future projects.

West of Duddon Sands also aids from having primary use of two of the world's most technically progressive connections vessels which is the Pacific Orca and the Sea Installer, employed successfully in tandem to construct basics and install turbines.

8.9. RENEWABLE ENERGY IN THE TRANSPORT SECTOR: FUEL CHANGE AND ELECTRIFICATION

Due to incomplete solutions, the conveyance sector is the most problematic sector to supernumerary fossil fuels for renewables. Energy competence, via lesser and cost-effective vehicles, transport mode vicissitudes from separate to public transport and the outline of e-mobility covers the way cumulative renewable energy source in the conveyance subdivision.

E-mobility, although does not only consists of electric vehicles, but also electronic powered trains, subways, buses, and taxis. All over the globe, there are incomplete but cumulative creativities to link electronic transport systems with renewable energy, chiefly at the city and local levels.

8.9.1. Bioenergy for the Transport Energy

In the year 2013, the liquid biofuels met around 2.3% of the total demand of transport fuel demand. Along with the small but on the same side growing use by the aviation and marine sectors. By the end of the year 2013, the usage of liquid biofuels has grown by a factor of three as compared to the year 2004 levels.

It is to be noted that the global production of fuel ethanol grew progressively from 32.6 billion liters in the year 2004 to 116.5 billion liters in the year 2013. From the earlier four years, the annual production of liquid biofuels has continued to be unevenly equal to 2010 levels.

The new facilities were also started. New ethanol and biodiesel production facilities were opened in the current years. Even though there were many ethanol plants that remained to operate below the capacity.

8.10. EVOLUTION OF GLOBAL INVESTMENT IN RENEWABLE ENERGY: A DECADE RETROSPECTIVE

The worldwide new investment in the renewable power and fuels nurtured progressively from USD 39.5 billion in the year 2004 to USD 279.6 billion in the year 2011. On the other hand, the investment was reduced in the year 2012 from this all-time high down to USD 249.5 billion. And in addition to it, it was further decreased to USD 214.4 billion in the year 2013.

The total investment was reduced by 11% in the year 2012 and was down to 22% in the year 2013, as compared to the 2011 levels. There was a decline in the investment after several numbers of years of stable progress. This decline in the investment resulted from uncertainty over support policies in Europe and the United States. Also, this decline in the investment was from actual retroactive reductions in support.

Though, there are several numbers of factors that resulted in the decline of the investment. The decline in the investment was also resulted from a sharp decrease in technology costs, especially solar PV. Even as global investment in solar PV in the year 2013 reduced nearly 22% as compared to the 2012 levels. On the other hand, the new capacity installations improved by more than 32%.

These sharp decreases in the cost in the wind and solar PV sectors provided attractive offers of renewables in the new markets. It was mainly in the developing countries where there is strong requirement for new electricity generation capacities to fulfill the growing demand of the energy.

There were reports that included the undocumented investments in the hydropower projects that were larger than 50 MW in capacity. According to these reports, the total money that was invested as new investment in the projects of renewable power and fuels was not less than 250 billion of US Dollars.

The amount of the investment in the newly installed projects for the renewable energy was of the largest capacity as compared to the year 2013. In the year 2013, the capacity for the project for the renewable resources was near about 120 gigawatts. This was the highest ever investment for a project for renewable energy with a factor of four since the levels that were set in the year 2004.

In fact, there was an investment of near about one and a half trillion US Dollars in the field of renewable energy.

In the year 2004, the commercial banks were planning to enter the explorable sector of renewable energy sector and start some investments in this totally new sector. This was the same time when the investment in the sector of renewable sector was considered as unsure and too risky by the majority of financiers and other investors.

The region of renewable energy was considered too risky because the worth of this sector was not proven. In the present times, there are many different institutions like the pension funds, major insurance companies, major corporations that have no connections from the energy industry, and many others have partnered with the commercial banks.

These institutions are looking for stable, long-term returns from their investments in the sector of renewable energy.

With the passage of time, the sector of the renewable energy is turning out to be equally competitive in the economic aspect. Even the investors are recognizing the enhancing worth of this sector. For the further development of this sector, it will be important to improve the designs and policies of effective financing tools to cover the initial investment costs in these projects.

There have been several innovations in the financial areas that have been observed since the year 2004. In the recent times, the investment firms have begun using the tools for better and stable returns on the projects for the renewable energy. There are many tools in the sector of renewable energy that are being used in the different countries of the world.

For instance, the use of the Sustainable Yield Bonds in the United States, Green Bonds in France and the United States, and the Renewable Financing Company Bonds in the United Kingdom help the investors in gaining a secure return from this sector. There is a scope for further innovations in the sector of investment for the renewable sources.

These innovations might vary from the crowdfunding to the modern ownership models. These new innovations are helping in turning the policies into reality for the individuals and communities and encourages the investments in the sector of renewable energy.

8.11. GLOBAL EXPANSION OF RENEWABLE ENERGY

According to some observations by the year 2014, it was found that there were at least 144 countries across the world that has set the targets for the renewable sources of energy. Along with this, there are near about 138 countries that already had policies supporting the use of renewable energy. This numbers increased with a great rate as there were near about 48 countries that were supporting the policies of the renewable energy in the year 2004.

The countries with developing and evolving economies have initiated the expansion in the recent years. Surprisingly, these countries are now responsible for near about 95% of the countries all across the world with the support policies for the renewable energy. This number has gone up from 15 in the year 2005 to the present situation.

In fact, most of the renewable energy policies are applied after keeping the focus on the power sector. There are many parameters like the feed-in tariffs and renewable portfolio standards that are the most popular instruments for the successful enforcement of these technologies. Similarly, there are methods like public competitive bidding or tendering that have gained further prominence in the acquisition of the tenders for the projects of renewable energy.

In fact, the number of the countries that favor the public auctions of the tenders has seen a steep alleviation. In the year 2009, there were only nine countries that reportedly accepted the method of the public auction. This was raised to near about fifty-five countries across the world as registered in the year 2014.

There are targets and guidelines that constantly support the renewable energy and have observed a gradual rise in the heating and cooling of the are steadily increasing. The mechanisms of the policy are continually evolving. These mechanisms have been distinguished by the continual usage of the technology.

The development of the feed-in policies in many countries has led to the evaluation of the methods of premium payments in the power sector. These companies and organizations have continued the use of such policies in the heating sector. Recently, there are many innovative policies that are emerging to the scenario.

These policies are very accommodating in the management of the integration of high shares of renewable electricity into existing power systems. These high shares include the support for storage of the produced energy, the management of the demand-side, and the encouraged usage of the smart grid technologies.

There are many developed countries that have been the initiators of the usage of such policies in the sector of renewable energy. Germany was one of the first countries across the world that established and applied the modern methods of renewable power generation. In fact, Germany especially took the interest in the development in the methods of the resources like onshore wind and solar PV. Between the year 2000 and the year 2013, the Ministry for Environment in Germany initiated a detailed and one-of-its-kind effort to observe the effects of the execution of the renewable energy sources.

8.11.1. Greenhouse Gas Emissions

One of the major effects of the usage of the renewable sources of energy is the cut-down in the emissions of the GHG. The enhanced market that was created for the renewable energy deployment decreased the rate of the emissions of the greenhouse gas as well as dust emissions throughout Germany.

In fact, the overall rate of the greenhouse gas emission has been reduced and has added to near about 148 million tons for the year 2012.

From other reports, it was observed that the biomass sector contributed the largest share of the percentage for the offset of carbon dioxide. This led to the decreasing near about 40% of the 148 million tons of carbon dioxide that was equivalent in the year 2013.

On the other hand, the sectors of the wind PV and solar PV each reduced an equivalent of near about 28% and 16% of the greenhouse gas emissions respectively. On the similar track, the other resources like the hydropower, biofuels, solar thermal and geothermal heating cumulatively reduced the emissions of the GHG by nearly 17%.

8.11.2. Fossil Fuel Savings

The usage of the renewable energy source for power generation, heat supply, or as a transport fuel helps in decreasing the fossil fuel demand.

It is benefited for the energy resource-poor country such as Germany has reduced the imports of fossil fuel.

According to the reports, in the year 2012 alone, near about 40 million tons of coal, 12 billion cubic meters of gas and 8 million liters of oil, diesel, and gasoline were saved in Germany.

8.11.3. Fossil Fuel Import Costs

There are many countries in Europe that imports most of its oil, gas, and coal from the other countries of the world. But, with the enhanced usage of the renewable energy in the energy sector, the total cost for the imported fossil fuels has seen a gradual decrease in the last ten years.

In the year of 2011 evaded fossil fuel introductions totaled up to Euro 6.6 billion, with the utmost savings in the power field. By the year 2012, the fossil fuel cost value savings were just over Euro 10 billion with the heating subdivision getting past the power sector by Euro 1 billion.

8.11.4. Cost of Renewable Energy

Let's take example of Germany's ground-breaking role in the growth and placement of renewable energy knowledges, which leads to a global drop in reliable cots. Cost for renewable energy technological innovations which includes onshore wind, solar collectors, bioenergy were ceased to be plummet suggestively.

Solar PV module prices plummeted by one order of scale; prices in the year 1980 were an average of united states dollar by 30/with peak plummeting by 97% to United States cent by 90/with peak in the year 2013. In Germany, cost discounts were attained chiefly due to the Feed-in Law.

8.11.5. Investments in New Technologies

Hiked investment in brand new renewables interprets into innovative jobs in the renewable energy industry. In accordance to the German government, the overall investment in the year 2012 renewable energy technologies was up to Euro 19.5 billion.

8.12. GLOBAL ENERGY CHALLENGES

The fusion of the energy market all over the world in the future will totally depend on the further technological advances. The most specific area in these advancements will be in the area of the development of the energy sources that emit low amount of carbon. However, it is quite evident that the energy sector is tremendously concentrated in the capital investment. There is always a need of the stable and a foreseeable regulatory guideline in the evolving sector of the energy market. This will be required to help the investors gain some returns or incentives from their investments. These incentives could prove as the capital for the new energy projects.

Especially in Europe, it is becoming gradually evident that the decisions that affect the energy sector are going to play an important role for the success of the economies of the countries that adapt these technologies in the years to come.

According to the chairman of The World Economic Forum's Energy Utilities Community, Mr. Ignacio Galán, since January 2014, there has been focus on the ever-growing problem of the regulatory doubt and the constantly changing priorities on the political front. This can be noticed in the implementation of the new policies of the taxation, the provision of the faltering subsidies and some other policies and guidelines that in many cases leave the utilities in a situation for the catch-all for the varied costs of the non-energy aspect. This ends up only as an addition to the bills the consumer has to pay.

There are many power companies that are gearing up to meet the technological obstacles of the rapid development of the technologies and the innovations in the new technologies. Some of these technologies, like the smart grids, are being used. Here, the policymakers are required to play a critical role and should know how to face and solve the regulatory inconsistencies.

According to the chairman of the World Economic Forum's Energy Utilities Community, Mr. Galán, "New plants can only be economically viable if they have a long-term, stable, and sufficient remuneration scheme, given the amount of investment they require. The right answer is to bet on the deployment of technologies that are mature enough to provide energy and environmental solutions at a reasonable cost and renewable energy is projected to be the fastest growing source of primary energy over the next decades."

He also said: "Subsidies to fossil fuel industries should stop so we can invest more in the fuels of the future. It also means avoiding massive subsidies for new technologies that are displacing efficient and relatively low carbon sources such as gas. A clear and long-term policy framework to enable investment decisions is required, overseen by a single European regulator to ensure consistency.”

REFERENCES

1. 10 Years of Renewable Energy Progress, (n.d.). [eBook] *France*, p. 4. Available at: http://www.ren21.net/Portals/0/documents/activities/Topical%20Reports/REN21_10yr.pdf (Accessed on 21 May 2019).
2. Burdett, R., (2015). *Technological Advancements and Their Impact on Renewable Energy Sector | The Renewable Energy Hub*. [online] Renewableenergyhub.co.uk. Available at: https://www.renewableenergyhub.co.uk/blog/technological-advancements-and-their-impact-on-renewable-energy-sector/ (Accessed on 21 May 2019).
3. Hub.globalccsinstitute.com (2015). *The Evolving Policy Landscape | Global CCS Institute*. [online] Available at: https://hub.globalccsinstitute.com/publications/renewables-2014-global-status-report/evolving-policy-landscape (Accessed on 21 May 2019).
4. Rangaraju, J., (2015). *Technology Advancements Accelerate Adoption of Renewable Energy*. [online] Electronic component news. Available at: https://www.ecnmag.com/blog/2015/10/technology-advancements-accelerate-adoption-renewable-energy (Accessed on 21 May 2019).
5. Reports.worldfinance.com (n.d.). *A World Finance Report: Technological Advances in Renewable Energy*. [online] Available at: http://reports.worldfinance.com/technological-advances-in-renewable-energy/ (Accessed on 21 May 2019).
6. Shwartz, M., (2017). *Future of Energy: Renewables | Stanford News*. [online] Stanford news. Available at: https://news.stanford.edu/2017/10/05/future-energy-renewables/ (Accessed on 21 May 2019).
7. Temple, J., (2019). *The Biggest Clean Energy Advances in 2016*. [online] MIT technology review. Available at: https://www.technologyreview.com/s/603275/the-biggest-clean-energy-advances-in-2016/ (Accessed on 21 May 2019).

INDEX

P

R

S

T

W